Food Supply Chain Man

Edited by

Michael A. Bourlakis and Paul W.H. Weightman

School of Agriculture, Food and Rural Development,
University of Newcastle upon Tyne, UK

Editorial Offices:
Blackwell Publishing Ltd, 9600 Garsington Road, Oxford OX4 2DQ, UK
Tel: +44 (0)1865 776868
Iowa State Press, a Blackwell Publishing Company, 2121 State Avenue, Ames, Iowa 50014-8300, USA
Tel: +1 515 292 0140
Blackwell Publishing Asia Ltd, 550 Swanston Street, Carlton, Victoria 3053, Australia
Tel: +61 (0)3 8359 1011

First published 2004 by Blackwell Publishing Ltd

Library of Congress Cataloging-in-Publication Data
Food supply chain management/edited by Michael A. Bourlakis and Paul W.H. Weightman.
p. cm.
Includes bibliographical references and index.
ISBN 1-4051-0168-7 (softcover : alk. paper)
1. Farm produce – Great Britain – Marketing. 2. Food industry and trade – Great Britain. 3. Marketing channels – Great Britain. I. Bourlakis, Michael A. II. Weightman, Paul W.H.

HD9011.5.F66 2003
381′.45664′00941–dc22
2003052363

ISBN 1-4051-0168-7

A catalogue record for this title is available from the British Library

Set in 10/13pt Times
by Graphicraft Limited, Hong Kong
Printed and bound in Great Britain using acid-free paper
by MPG Books Ltd, Bodmin, Cornwall

For further information on Blackwell Publishing, visit our website:
www.blackwellpublishing.com

Contents

Preface

Food is always a matter of interest, a means of providing energy, the raw material that builds and maintains our well-being, a defence against illness, a pleasure to consume when well prepared and presented, a basis for social interaction and enjoyment at home, in a restaurant, canteen or perhaps even hospital or school. Our ancient ancestors spent more than half of their waking hours searching for, gathering, hunting and preparing food.

Clearly we have progressed. For example, television and magazines have large audiences eager for entertainment and instruction on methods of food preparation and presentation, of novelties, of variations on a theme, exotic and ethnic foods or perhaps the simple basics such as ensuring an egg is fresh, boiled, fried or poached to perfection. Yet, whoever we may be, we rely on a secure supply of food, and that is more or less taken for granted in richer countries. Consumers' concerns are price, ease of acquisition and preparation, and quality as measured by preferences for colour, taste and consistency.

By and large, all these are met for the majority of people in developed countries for most of the time. We live at a time of abundantly supplied shops, stores, supermarkets and fast food outlets. Supporting all of these is a supply chain, a series of links and inter-dependencies from farm to fork, from plough to plate. It is the behind-the-scenes production and preparation that delivers hour by hour, week by week, through almost seamless seasons, providing us with food at a price we like and in a place we want it. The food supply chain is a managed process, a combination of knowledge and skills, spanning electronics, biology and the social sciences of economics, human behaviour, psychology, and more. All are set in a legal framework of minimum standards and basic rules.

The food supply chain embraces a wide range of disciplines. This book brings together the most important of them and aims to provide an understanding of the chain, to support those who manage parts of the chain and to enhance the development of research activities in the discipline. It will therefore be of interest to students, managers, researchers and policy decision-makers with an interest in food supply chain management.

The book follows a 'farm to fork' structure. Each chapter starts with aims and an introduction, and concludes with study questions that students in particular may find useful. The editors introduce the food chain environment, setting the scene by describing its major parameters and descriptors. Consumer satisfaction is the main objective of every chain, and two chapters by David Marshall, Sharron Kuznesof and Mary Brennan relate to this. They discuss, among other things, the product choice process, purchasing behaviour and aspects of perceptions of risk concerning food safety. The procurement function is introduced in a further chapter by Johanne Allinson.

Aspects of crop and livestock production are addressed by Stephen Wilcockson and David Harvey. Their chapters illustrate the critical role of agriculture and primary producer within the food supply chain that has been neglected by many books.

Food manufacturing, the processors, assemblers and preparers of near ready or table ready food are described by David Hughes, while wholesalers, retailers and caterers are discussed by John Dawson. Networks and alliances between links in the food chain are illustrated by Rachel Duffy and Andrew Fearne, and Mark Francis describes new product development and the use of information technology by Tesco, the UK's leading food retailer.

In Chapter 11, Alan McKinnon stresses the pivotal role of third party logistics firms in the modern food supply chain, while temperature-controlled supply chains are addressed by David Smith and Leigh Sparks in the following chapter.

Chapter 13 has been written by a research team led by Carlo Leifert and contributes to our understanding of the organic food supply chain, which is based on a traditional way of growing crops without the support of modern fertilisers, herbicides and pesticides. Following that, James Stock discusses the US food supply chain, since many of the innovations to the UK and European supply chains (e.g. efficient consumer response) originated on the west side of the Atlantic.

In the last chapter, Costantine and Michael Bourlakis provide a synoptic view of the trends within the food supply chain's dynamic system and the likely direction of its future management.

It is our aim to promote knowledge and understanding of the UK food supply chain; however, we anticipate that the book will prove useful to readers based in other countries in Europe and around the world. We are delighted to be able to include knowledge and expertise from contributors based primarily in leading UK universities. Many of the contributors are currently working at the University of Newcastle upon Tyne, or are either graduates or former staff members. This is hardly surprising, since Newcastle was the first English university to introduce a Chair of Agriculture, in 1892, and the first UK university to introduce a professorship in food marketing in 1963.

The food industry has developed massively, from the fundamentals of production, harvesting and storage, to a situation where productivity is enhanced through genetics, chemistry, mechanisation and management. There was a time in the memory of at least one of the authors when the UK struggled to match supply with demand, but food production and supply have moved on, so that quality, safety, presentation and ease of use are at the top of the list of consumers' demands. This book will provide an understanding of how the supply chain works and how it will develop to meet these needs.

M. Bourlakis and P. Weightman

Forewords

There can be no doubt that supply chain management has moved much higher up the corporate agenda in recent years. It is now widely recognised that the supply chain holds the key in many industries to cost reduction as well as service enhancement. There are several reasons for this belated recognition of the importance of supply chain management. First, the business model in the past was often based upon a philosophy of vertical integration whereby upstream and downstream facilities and activities were owned and managed by one organisation. Today the pendulum has swung the other way. Now we talk about outsourcing all activities other than our core business. The extent of this outsourcing in some instances is such that we should not talk of supply chains but rather supply 'networks'.

A second driver of supply chain complexity is the globalisation of industry. Not long ago the norm was 'local for local' manufacturing and distribution. Now global sourcing and focused manufacturing in fewer but bigger facilities is more frequently the case. Whilst the motivation for global sourcing and focused manufacturing is clearly the search for lower unit cost, it can be argued that this view of cost is too narrow and that the overall impact on the business could be to extend lead-times and to increase vulnerability to disruption.

A third factor that is particularly relevant to the food industry is the growing demands placed upon suppliers by ever more powerful retailers. Retail concentration is now a fact of life in many markets and is increasing as we see the emergence of global retailers. Their demands for just-in-time delivery, for higher product quality and tailored logistics solutions means that suppliers must review their supply chain strategies.

Against this backdrop of dramatic change in the supply chain landscape it is particularly timely that this collection of insights into the food chain is being published. There is no aspect of everyday life that is more critical today than the means by which the world is fed. Surprisingly, for something as critical as food production and distribution, this subject has had very little exposure outside the small group of specialists who work in this area. I am convinced that this neglect is about to change as we all come to recognise the fragility of the systems that endeavour to ensure that we all can continue to enjoy the availability, convenience and experience that has come to typify today's food supply.

Professor Martin Christopher
Director, Cranfield Centre for Logistics and Supply Chain Management

* * *

The UK food and grocery supply chain employs over 3.2 million people, 16% of the total UK workforce. It is responsible for over £125 billion of consumer expenditure and contributes to over 8% of GDP. The UK food and grocery market is also one of the most competitive and customer focused in the world. Strong competition, coupled with a passion to delight the consumer, inevitably leads to intense pressure on all parts of the supply chain to perform more efficiently, drive out unnecessary cost, add value, innovate and offer competitive prices. The pressures, issues and indeed the opportunities vary, and some overlap, depending on where in the supply chain companies are positioned.

For UK farmers, many of whom have unquestionably suffered considerable erosion of their incomes in recent years, one of the key challenges is the gradual shift away from production support mechanisms. Those who can best meet the needs of the market will be in a better position to prosper in the future. The challenge for many individual farm businesses and farming groups is to decide if they are in the business of producing commodities, or if they are going to differentiate their production, i.e. make their goods 'special' and add value for their customers. For those who remain as commodity producers, the challenge is to be efficient, low-cost and large scale. There is a school of thought which says that, if farmers get caught between these two strategies, they may well find the way forward difficult.

For those farmers who have recognised the need to add value, local sourcing has been put forward as a possible route to prosperity. The market for local food is growing and retailers are responding positively to these customer demands. Yet it won't be the solution for everyone. Like organics, the market is bound to be niche at best.

Moving down the supply chain to processors and manufacturers the issues are, to some extent, similar. UK manufacturers have excelled in producing new and interesting products which meet consumers' needs. Innovation in chilled ready meals is a good example of this. However, the need to be highly price competitive has, to a certain extent, driven some food manufacturing to other countries. Certain areas of the UK food manufacturing sector now find it difficult to compete, but not because they are inefficient or lack innovation, far from it. One of the major difficulties is the lower cost of labour elsewhere, particularly when we look beyond Europe. A good example of this is the poultry industry where the demand from consumers has grown fantastically. Yet two-thirds of all the poultry consumed in the UK is now imported from countries such as Brazil, Thailand and China, despite the fact that poultry integrators have developed super-efficient, totally vertically integrated chains, often starting from the cereal crops that form the key ingredient in the chicken's feed, through to manufacturing processed, highly innovative products for retailers and caterers. With the UK supply chain so reliant on manual labour it is perhaps unsurprising that competing with imports is at best challenging, despite all that companies have done to become more efficient and consumer facing.

Our food retailers face challenges from many quarters. One is the unstoppable growth in consumer demand for convenience, which can often manifest itself in a decision to eat out, rather than cook at home. Retailers in the UK, whether the larger supermarket chains or convenience store chains, have, for some time, been the model emulated by other countries. Most such retailers are highly focused on giving their customers what they want, but with the growth in eating out, the competition is coming not just from each other but from new quarters – the take away, restaurants and hotels.

The response has been swift and creative. It has included the development of in-store home meal replacement centres, food-to-go, in-store dining opportunities, improvements in chilled ready meals and even drive-through takeaways. It has also included innovative category management and merchandising developments and the use of food service brands in food retail environments.

The food service sector is growing rapidly, so the supply chain challenges here are quite different. The sector is serviced by a small number of large players who operate restaurant and hotel chains and who service the 'cost sector' (e.g. schools, prisons and public services), and a very large number of small chains and independents.

One of the key differences between the food retail and food service sectors is the efficiency of their supply chains. Taking the sectors as a whole (whilst appreciating there will be wide variations between individual players), food service is characterised by a certain degree of supply chain inefficiency. This presents a challenge for suppliers who may wish to take advantage of this growing part of the market – how can they efficiently supply a large number of small operators with no centralised distribution?

So where does this lead us? The common thread to success, no matter in which part of the supply chain a company operates, is to understand the needs of both the next customer in the chain and the end consumer. The pace of change in consumer needs is accelerating. Those who can adapt quickly and keep up with consumers stand the best chance of prospering. This means investing in good quality information about the consumer. Knowing what the consumer and your customers require should be the main pillar in any consumer facing industry. The added challenge is to meet consumer needs better, faster and at less cost and this is where the Institute of Grocery Distribution (IGD), a leading source of information, research and education for the food and grocery industry, plays its part.

Being aware of the challenges by researching and disseminating information for the market and developing best practice is only half the solution. Making it happen is the other half. That requires the right people who understand how to manage effectively and are prepared to take risks. The food industry is an exciting, rewarding and dynamic place to work and needs bright, energetic people to take it forward. Through their research and teaching, universities and colleges play a vital part in developing this talent. I am confident that this book will help this process.

Richard Hutchins
Business Director
Institute of Grocery Distribution
March 2003

Contributors

Johanne Allinson is a research associate in the School of Agriculture, Food and Rural Development at the University of Newcastle upon Tyne. Since joining the University in 2001 she has worked on a number of agro-food and rural development research projects, including a study of the UK fat lamb chain following the footh and mouth outbreak in 2001, and an evaluation of the impact of foot and mouth disease on food consumers. She has previously conducted research into the Cumbrian agro-food sector, including a study of potential infrastructure development opportunities for that sector with the Centre for Regional Economic Development at the University of Northumbria. Her core research interests include the structure, form and nature of agro-food supply organisation and regulation and the power relations that characterise them, and the concept of regional or local rural development. She previously lectured on rural development at the University of Northumbria and was a rural development officer at Durham County Council.

Constantine A. Bourlakis is a senior lecturer in microeconomics at the Department of Business Administration, Athens University of Economics and Business, Greece. He has a degree in economics and statistics from the Athens University of Economics and Business, an MA in economic studies from the University of Newcastle upon Tyne, and gained his PhD from the University of East Anglia. He has worked as a research associate at the University of Warwick and as a lecturer at the University of Leeds, the University of Edinburgh and the University of Leicester. Constantine's research interests lie in the broad areas of industrial economics and business strategy.

Michael A. Bourlakis is a lecturer in food marketing at the School of Agriculture, Food and Rural Development at the University of Newcastle upon Tyne. He graduated in business administration from the Athens University of Economics and Business and obtained both his MBA and PhD degrees from the University of Edinburgh. Michael worked as a research associate at the Management Centre, University of Leicester, and at the Oxford Institute of Retail Management, Templeton College, University of Oxford. His research interests include retail logistics, agricultural logistics and food supply chain management and Michael has published in various logistics, supply chain management and marketing journals. He is also an editorial board member of *Supply Chain Management: An International Journal.*

Mary Brennan is a lecturer in food marketing in the School of Agriculture, Food and Rural Development, University of Newcastle upon Tyne. Her current research interests include the communication of scientific messages to the public, in particular food risk and uncertainty information, public involvement in science and technology, for example, in the development of novel foods and technologies, and the use and development of qualitative research methodologies in applied food policy research.

Gillian Butler is a graduate of Edinburgh University and works at the School of Agriculture, Food and Rural Development at the University of Newcastle upon Tyne. She is responsible for setting up a consultancy centre (within the Tesco Centre for Organic Agriculture) to support the organic red meat industry in the north of England and southern Scotland, and teaches undergraduate students in agriculture. Her teaching duties include the environmental impact of farming, beef and sheep production, animal nutrition and aspects of organic livestock management. Gillian also works as an agricultural consultant, offering advice to conventional and organic farmers on ruminant production, with particular reference to beef and sheep nutrition and grassland management. This is a role she has carried out for the last 25 years, having worked with the Agricultural Development and Advisory Service (ADAS) for 17 years before turning freelance in 1994.

John Dawson is Professor of Marketing at the University of Edinburgh, Visiting Professor at ESADE and Distinguished Professor at the University of Marketing and Distribution Sciences, Kobe. He has Bachelor and Master degrees from London University and a PhD from the University of Nottingham. His activities in research, management development and teaching focus on retail strategy, marketing and relationships with suppliers. John publishes, consults and makes presentations widely in Europe and Japan. In the last few years he has held positions as Visiting Professor at Bocconi, Milan, the University of South Africa at Pretoria, Saitama University and the University of Marketing and Distribution, Kobe. He is chairman of a small retail company that operates shops in museums in Scotland. He is currently working on a joint project with the University of Stirling and Glasgow Caledonian University to create a Centre for the Study of Retailing in Scotland, and is involved in a major study with scholars in Japan, the UK and France, looking at the internationalisation of retailing.

Rachel Duffy is a research associate at the Centre for Food Chain Research, Imperial College at Wye, and was recently awarded her PhD which investigated the performance implications of supply chain partnerships for suppliers in the fresh produce sector. Her research interests are varied and have included the behavioural dimensions of supply chain relationships such as interdependence, collaborative activity, trust, relational norms, commitment and conflict resolution. Recent research projects include an investigation of the fairness and integrity of trading relationships between supermarkets and suppliers in response to the findings of the Competition Commission report published in October 2000.

Andrew Fearne is Director of the Centre for Food Chain Research at Imperial College, London. He graduated in French and economics from Kingston University and gained his PhD in agricultural economics from the University of Newcastle upon Tyne. After a brief spell with the National Farmers' Union, Andrew returned to Newcastle where he spent six years as a lecturer in agricultural commodity marketing. In 1994 he moved to Wye (now the Department of Agricultural Sciences at the University of London), as a senior lecturer in food industry management. He has two specific research and consultancy interests, supply chain management and consumer behaviour, with particular emphasis on the creation, management and exploitation of value-added in the dairy, livestock and fresh produce industries. Andrew is editor of *Supply Chain Management: An International Journal* which addresses both practical and research issues concerned with the links in the food supply chain, from the breeders, primary producers, manufacturers, retailers and caterers to the final consumer.

Mark Francis followed a successful career in the software development industry where he spent ten years working as a software engineer, business systems' analyst and software support specialist. He has a degree in computer studies and an MBA from Cardiff Business School. Mark joined the staff at Cardiff Business School in 1996 as a senior research associate on the Supply Chain Development Programme (SCDP). He has worked on a number of applied research projects in the field of operations and supply chain management and he now specialises in the subjects of product development practice and process design in the food industry, which form the content of his recently completed PhD. He is a co-director of the recently created Agri-Food Process Innovation Research Unit, and is currently working on the DTI-funded project, Red Meat Industry Forum: Value Chain Analysis. This is in collaboration with the MLC, NFU and the new Food Chain Centre at the Institute of Grocery Distribution (IGD).

David Harvey has a degree in agriculture (agricultural economics) from the University of Newcastle upon Tyne, and an MA and PhD in agricultural economics from the University of Manchester. Between 1974 and 1979 he was a research economist with the Canadian Department of Agriculture, and latterly with the Economic Research Council in Ottawa, specialising in grains and grain transportation policies. He returned to Newcastle as a lecturer in 1979, moving to a professorship at the University of Reading in 1985, and moved back to Newcastle upon Tyne in 1987 to take up his present position. His research interests focus on agricultural policy analysis and the policy process, the European Common Agricultural Policy, international policy and rural and environmental policies. He has been involved with modelling the agricultural/environmental interface in UK agriculture, with major research awards from the Economic and Social Research Council and the Natural Environment Research Council, and also with contracts with the MAFF and DoE. He has also been involved with research on possible alternative policies consistent with the objectives of the GATT Uruguay Round with the International Agricultural Trade Research Consortium, and is a co-designer of the Producers' Entitlement Guarantee (PEG) proposal with colleagues at Cornell University. More recently, he and his colleagues have been involved in analysis and advice on Central European liberalisation and transition policies. His present research interest is focused on developing a multi-disciplinary evolutionary approach to policy and management.

David Hughes is Professor of Food Marketing at the Centre for Food Chain Research, Imperial College, London, and Visiting Professor at the Royal Agricultural College, UK. David is a regular speaker at international conferences and seminars on global food industry issues, and is a proponent of building vertical alliances between key food chain members in the food industry – farmers, manufacturers, retailers and food services. He has lived and worked in Europe, North America, the Caribbean, Africa and south-east Asia and, currently, spends two months a year working with food industry participants in Australasia. David is an international advisory board member with food organisations in the UK, Holland, USA, Canada and New Zealand; a co-owner of a fresh produce company in the USA and a non-executive director of KG Fruits, a UK farmer-owned soft fruit business. He works closely with senior management of food firms on business strategy development and with governments on food policy formulation.

Sharron Kuznesof is a lecturer in food consumer behaviour and food policy in the School of Agriculture, Food and Rural Development, University of Newcastle upon Tyne. Her current research interests include public attitudes to food risks and food safety, the communication of food risks and food risk uncertainty, and the public acceptance of novel foods and technologies, in particular gene technology and genetically modified foods.

Carlo Leifert is Professor for Ecological Agriculture and Director of the Tesco Centre for Organic Agriculture in the School of Agriculture, Food and Rural Development at the University of Newcastle upon Tyne. He is also a director of the Organic Glasshouse Research Centre (OGRC) which is part of the Stockbridge Technology Centre (STC) in the UK. Over the last six years his applied research group has focused on improving agronomy, efficiency and food quality and safety in organic production systems. This work was carried out in close collaboration with the supermarket Tesco and organic producers and processors, and focused mainly on quantifying interactions between human nutrition and health, and the soil environment and crop health. Previous appointments include Director of the Aberdeen University Centre for Organic Agriculture (AUCOA), Technical Director at Howegarden Ltd and R&D Manager at Neo Plants Ltd. His applied research work has concentrated on the development of HACCP-based quality assurance systems for organic and conventional horticultural production systems, food processing and plant biotechnology facilities.

David Marshall is a senior lecturer in marketing and consumer behaviour at the University of Edinburgh. His main research interests relate to consumption behaviour. They include understanding social aspects of eating and the role of meals in food choice, attitudes and behaviour in relation to healthy eating, the impact of marriage and co-habitation on young couples' eating habits, and food availability. He has contributed to a number of books and edited *Food Choice and the Consumer*, a collection of papers on consumer issues related to food provisioning. He has published in several journals including *International Review of Retail, Distribution and Consumer Research, Journal of Marketing Management, Appetite, Food Quality and Preference, The British Food Journal, The Sociological Review, Health Education Bulletin* and the *Health Education Journal.*

Alan McKinnon is Director of the Logistics Research Centre and a professor at the School of Management at Heriot-Watt University, Edinburgh. Alan has been researching and teaching in the field of freight transport for 25 years and has published widely on the subject. He has been an advisor to several UK government departments and consultant to numerous public and private sector organisations on a variety of logistics and transport issues. He is a Fellow of the Institute of Logistics and Transport and a founder member of its Logistics Research Network, set up to promote research in UK universities and colleges on logistics and freight transport.

Helen Newton is a faculty information support officer at University College, London. Helen graduated in animal and plant biology at Sheffield University and obtained her MSc in crop protection from Harper Adams Agricultural College. Helen worked as a sales agronomist and then as a demonstrator in agronomy at the University of Newcastle upon Tyne, teaching crop protection and production and IT for agriculture. She was involved in a research project detailing the future need for organic seed in Europe for the Tesco Centre for Organic Agriculture and authored a regular column on organics in the local newspaper. Current research interests include IT competence and online teaching materials in staff development.

David L.G. Smith is Principal Consultant at DLG Logistics. He is an experienced senior logistics operator, having worked for Tesco for nearly 15 years. During a period of major innovation and change in the Tesco distribution networks, he held positions as Head of Primary Distribution, Head of Composite Distribution and Divisional Director for Contract Distribution. In 1998 he was seconded to the Department for Transport to advise on sustainable distribution strategy and best practice from the logistics industry. At present he is working on several transport projects for government, including the national food fleet survey and the fuel economy advisors' scheme. David obtained his MBA in 1992 at the University of Stirling, specialising in retail logistics and temperature controlled supply chains, subjects on which he has written and lectured to retailers, suppliers, manufacturers, logistics service providers and universities. He is currently co-writing a specialist book on the subject, to be published in 2003.

Leigh Sparks is Professor of Retail Studies at the University of Stirling, UK. His research concentrates on aspects of retailing, with particular emphasis on structural and spatial change, and has been funded by major businesses, research councils and governments. Leigh has published widely in leading retailing and marketing journals. He graduated from Christ's College, University of Cambridge, and was a postgraduate and research fellow at St. David's University College, Lampeter. Since arriving at Stirling he has also been a lecturer and senior lecturer. Previously he was the head of the Department of Marketing, the Director of the Institute for Retail Studies and the Dean of the Faculty of Management. Leigh has been a Visiting Professor at the College of Human Sciences at Florida State University in Tallahassee. He is currently Director of the SHEFC-funded Centre for the Study of Retailing in Scotland. He is co-editor of *The International Review of Retail, Distribution and Consumer Research.*

James Stock is Professor of Marketing and Logistics and holds BS and MBA degrees from the University of Miami, Florida, and a PhD from Ohio State University. He has held faculty positions at the University of Notre Dame, University of Oklahoma, Air Force Institute of Technology and Michigan State University, before joining the University of South Florida. He is currently Editor of the *International Journal of Physical Distribution and Logistics Management* and was formerly the Managing Editor of *Logistics Spectrum.* James has published in numerous academic and professional journals and is the author or co-author of over ninety publications including books, monographs, articles and proceedings papers. He is co-author of *Strategic Logistics Management* (4th edn) and *Fundamentals of Logistics Management*, and author of *Development and Implementation of Reverse Logistics Programs and Reverse Logistics*, all published by the Council of Logistics Management. In 1988, James received the Armitage Medal from SOLE, The International Society of Logistics, in recognition of his scholarly contributions to the logistics discipline. His areas of interest include reverse logistics, the marketing/logistics interface and supply chain management.

Paul Weightman is a lecturer at the University of Newcastle upon Tyne. He has a first degree in agricultural science from Durham University, an MSc in agricultural economics from Cambridge University, and a PhD in farm and business management from Cornell University in the USA. He has taught and consulted widely, for the US Department of Agriculture on implications for UK agriculture of EU enlargement in the 1970s, the Overseas Development Administration in Malawi on farm business surveys and on management in central Europe over four years of transition from 1989. He is a part-time arable farmer in Cleveland, North Yorkshire, and, since 1996, a director of a Portuguese agribusiness supplying winter salads to UK multiple retailers and floral decoratives to continental European and Scandinavian markets. His current research interests are quality assurance and traceability.

Steve Wilcockson is a lecturer in crop production in the School of Agriculture, Food and Rural Development, Faculty of Science, Agriculture and Engineering at the

University of Newcastle upon Tyne. As well as teaching and undertaking research, he has been Admissions Tutor for the degree programme in agriculture for many years and also the Degree Programme Director. Steve's teaching and research interests focus on non-combinable crops, particularly potatoes, sugar beet and field scale vegetables in both conventional and organic production systems. His research covers agronomic and physiological factors influencing crop growth, yield and quality for different market outlets. His teaching subjects include production in the field, processing and marketing, through to the final consumer. Steve is also an assistant co-director of the Tesco Centre for Organic Agriculture at the University of Newcastle upon Tyne. The centre aims to support expansion of organic production in the UK, improve yield in organic production and processing, and transfer appropriate 'organic' strategies to conventional systems.

Chapter 1

Introduction to the UK Food Supply Chain

M. Bourlakis & P. Weightman

Objectives

(1) To provide the contextual setting for succeeding chapters and particularly the dynamics of the UK food industry.
(2) To provide an insight into the factors that have influenced change and development of the food supply chain.
(3) To illustrate the value added provided by each supply chain activity.

Introduction

Food is one of the pillars upon which society is built. It is fundamental to health, happiness and political stability. There will always be challenges in how society develops, how it changes, in the relationships between peoples, classes and nations. But when there is a problem with the food supply chain's ability to deliver, then it becomes the first priority for the security and personal safety of individuals and the nation.

This book is written at a time when, in Europe, there is an abundance in the supply of food of all types and at prices which are probably the lowest they have ever been when compared to average wage. This success is a product of foresight, effort, planning, competencies and skills and many interdependencies of individuals, businesses, organisations, governments and regulatory authorities that span many disciplines. The book will outline key aspects of the management of the food supply chain.

Food supply chain management is developing as a research discipline in its own right. It spans local, regional, national and international arenas. It is an area that is bringing together a widening range of interests. The food supply chain has progressed from a series of shorter, independent transfers to more unified, coherent relationships between producers, processors, manufacturers and retailers. Hence, it is developing from a situation where relationships were categorised as trading, competitive, opportune and, at times, confrontational to one where they may be broadly described as longer term, larger scale, programmed, information-sharing, increasingly trusting, more transparent and with greater traceability, with defined responsibilities and yet retaining a competitive edge.

The impetus for change in the UK in the 1980s and 1990s has come primarily, but not exclusively, from the initiatives of the major food multiple retailers whose dominance in the market enables them to exercise considerable influence. They require the delivery of consistent quality at competitive prices and, following recent food scares, e.g. BSE and salmonella, food supply chain management aims to guarantee the provision of safe and healthy products that are fully traceable from 'farm to fork'.

Throughout this textbook, the reader should be aware of issues that are of concern to all parties, or links, in the chain. There are conflicts of interest, paradoxes, power struggles, uncertainties, elements of risk and, perhaps, injustices. It is an aim of the authors to illustrate and explain the food supply chain, to reveal its strengths and weaknesses, to facilitate systematic analysis and to further the understanding of its management.

The food supply chain management environment

Food supply chains operate in a complex, dynamic, time-critical environment where product integrity is vital. There must be a high degree of certainty that food will be of a certain quality. The next section outlines six key factors that play an influential role in the evolution and development of modern food supply chains.

Quality

Quality is the degree of congruence between customers' expectations and their realisation. A manager's task is to deliver the expectations of customers, workforces and suppliers. Public companies also have to deliver value for their shareholders.

It has been proposed that any activity that adds costs, for which buyers are not willing to pay, should be eliminated (Christopher, 1998). We believe this is not correct although, certainly at a fundamental level, price is the determinant of demand. In wealthier societies in general, consumers want quality and lower prices. Quality and its assurance have become important supply chain management tools. Nevertheless, there are difficulties and challenges. Many producers are separated from customers due to the length of the supply chain. For example, regional producers of mutton and lamb may be remote from their customers and, due to a reduction in the number of meat processing facilities and the distances over which the meat is transported, they are also separated from consumers.

Responses to these difficulties are discussed in later chapters. They include quality assurance schemes, production and manufacturing to retailers' protocols and the application of quality management systems and standards such as HACCP and ISO series. Those who are unable to adapt to the rigorous requirements of the modern supply chain, either through scale of enterprise or lack of knowledge or financial constraints, become non-competitive.

Technology

The food supply chain includes a technological dimension. Its evolution is made possible by the myriad of innovations and developments essential for its integrity, efficiency and ability to increase its productivity. These include accurate weighing, refrigeration, controlled atmospheric bacterial growth inhibition, pasteurisation, micro-element pollutant detection, bar coding, electronic recognition of packaging, the use of stabilisers, artificial insemination, embryo transplantation, precision seeding, environmentally and welfare friendly animal housing, and organic crop and animal production systems. These advances indicate some of the many applications of research in biology, genetics, biochemistry, chemical engineering, computing science and other disciplines to agriculture, food manufacture and distribution.

Logistics

Various researchers have argued that logistics is a key business process that provides increased customer satisfaction (see, for example, Bowersox & Closs, 1996). Harvey (1996) quoted Christopher as defining logistics as 'the process of strategically managing the procurement, movement and storage of materials, parts and finished inventory (and related information flows) through the organisation and its marketing channels in such a way that current and future profitability are maximised through cost-effective fulfilment of orders'.

It is useful to note that whilst logistics concerns primarily the processes of a single firm, supply chain management also encompasses the external environment of an organisation and subsequently includes the external flows of materials, information and revenue between various businesses (Bowersox & Closs, 1996). Furthermore, Christopher (1999) notes that supply chain management evolves around partnerships developed in the chain and is supported by information technology applications that co-ordinate information dissemination and sharing amongst the chain members.

Information technology

Information technology applications support the movement of products and product information dissemination in the food chain. For example, the identification of products with bar codes using optical (electronic) methods such as electronic point of sales (EPoS) (scanning) technology, is employed widely to identify product locations in retail warehouses and stores and to record product movement onto and off vehicles. An example of electronic transmission of information using standard protocols is electronic data interchange (EDI) based on a specific standard to co-ordinate various members of the supply chain (Fynes & Ennis, 1994). The Economist Intelligence Unit (1988) stated the benefits accruing from EDI to business as being faster trading cycles, improved inventory management, a reduction in working capital requirements, improved cash flow, security and error reduction, and acknowledged receipt of order and delivery. The combined use of EPoS and EDI facilitated the implementation of quick response technology in retail operations (Larson & Lusch, 1991) to provide higher sales and lower stock levels through rapid data collection and processing.

The use of the Internet in the food supply chain is a relatively recent innovation largely favoured by food suppliers and multiple retailers (Quarrie & Hobbs, 1997). For example, the major UK food multiple retailer J. Sainsbury introduced Xtra Trade, to allow suppliers to exchange supply chain information over the Internet; Tesco Information Exchange is similar (Anon., 1999). Although it appears to be just another EDI system, the major benefits from the use of the Internet are that it enables both retailers and suppliers to avoid excessive paperwork and each stage of the system is more visible.

The regulatory framework

The food supply chain is affected by the socio-political environment; the text will therefore refer to the regulatory framework defined by national and international law. This regulatory framework reflects increased consumer concerns about food safety, labelling and product traceability. Significant, but not all, relevant topics will be mentioned, with specific legislative measures being fully discussed in the following chapters.

Consumers

Consumers drive the supply chain; 'demand chain' would therefore be a more accurate description when the primary driving force in terms of type, volume, quality and value of food supplied is considered. A development in the food supply chain that advocates this view is efficient consumer response (ECR) where manufacturers, wholesalers and retailers work to meet consumer demands better and more efficiently (Fiddis, 1997).

Economic analysis is applied to obtain a better understanding of consumer behaviour and achieve extra insight into the strategic direction of change in food consumption. The basic characteristics of demand for goods and services can be explained on the basis of the price of almost all goods and the income of every purchaser. Price elasticity of demand measures the change in demand for goods when there is a change in price; cross price elasticity measures the change in demand for particular goods when the price of other goods changes – consumers will substitute one type of food for another. In both cases the elasticity measures the response in demand when income is assumed to be constant; the effect of income change is measured by the income elasticity of demand.

Table 1.1 Price and income elasticity of demand for meat and fish (Lechene, 1999).

	Price elasticity (1988–98)	Income elasticity (1997–99)
Beef and veal	−0.46	0.21
Mutton and lamb	−1.23	0.10
Pork	−0.94	0.15
Bacon and ham uncooked	−0.80	0.22
Bacon and ham cooked	−0.71	0.23
Poultry uncooked	−0.63	0.23
Poultry cooked	−0.90	0.30
All other meats and products	−0.26	0.21
Fresh fish	−0.75	0.41
Frozen fish and fish products	−0.38	0.11

Lechene (1999) calculates the price elasticity for meat and fish and their products in the period 1988–98 (Table 1.1): all are negative, and as price rises quantity demanded falls. Demand for mutton, lamb, pork, bacon, ham and cooked poultry had elasticity of around unity 1; demand for these goods falls and rises at about the same rate as price increases or decreases. Income elasticity of demand for all meats is low whilst for fresh fish it is slightly higher in 1997–99. A 1% rise in income only generates a 0.21% increase in demand for all other meats and products.

The food supply chain in the UK economy

Food and its supply are significant parts of every national economy. Typically, the food chain includes agriculture, food manufacturing, food and drink wholesaling and retailing, and the food catering and service sector (Figure 1.1). The food chain employs 12.5% of UK workers and accounts for 8% of the economy (DEFRA, 2002).

More specifically, the UK food and drink manufacturing sector is the second largest in terms of output when compared with any other manufacturing sector and employs around 0.5 million people; the value added by this sector was £19.9 billion in 2002 (Table 1.2) and its contribution to well-being, employment and exports is significant. UK farms supply about three-quarters of its raw materials. When returns to primary producers are low and there are declining levels of farm subsidy, there is a threat to UK-sourced food supply to the manufacturing sector.

Food manufacturers' contribution may be quantified in several ways: in terms of proportion of gross national product, proportion of consumer expenditure, numbers employed, impact of imports and exports of food on balance of payments, and value added. The diagrammatic presentation of the food supply chain shown in Figure 1.1 is translated into financial and employment parameters in Table 1.2, giving an insight into the structure of the chain and an indication of the contributions of sectors measured in value-added terms. Value added is value of sales less the cost of all inputs except wages and salaries; it is estimated by the Department of the Environment, Food and Rural Affairs (DEFRA) and is not a measure of

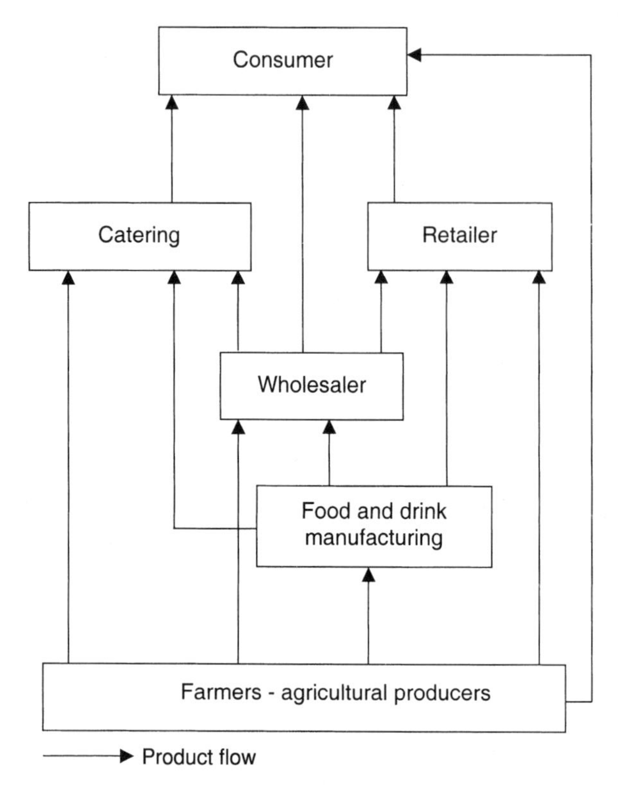

Figure 1.1 The UK food supply chain. The diagram excludes importing/exporting activities and focuses on product flow.

Table 1.2 The UK food supply chain in terms of value added and employment (DEFRA 2002).

	Value added		Employment (no. of employees)		Value added/employee	
	2002 (£bn)	1999 (£bn)	2002	1999	2002	1999
Retailers	17.0	12.5	1 110 000	947 000	15 454	13 199
Caterers	13.8	11.8	1 254 000	1 119 000	11 005	10 545
Wholesalers	5.5	4.6	192 000	220 000	28 645	20 909
Manufacturers	19.9	18.7	476 000	455 000	41 806	41 098
Farmers and producers	6.6	8.2	557 000	527 000	11 849	15 560

profitability because capital investment in each sector is not considered. Value added is an indication of the wealth created by employees and management through their labour and skills. Market conditions also influence value added; more favourable terms of trade created by excess supply and reduced prices of inputs into the manufacturing sector, for example, will increase the contribution.

Food manufacturers continue to deliver quality food consistently and efficiently. Some are household names, e.g. Heinz and Cadbury; others, such as Unilever and Nestlé, are recognised by their brands rather than the parent companies; Northern Foods and Geest are relatively unknown, but represent substantial businesses in prepared foods and meals for multiple retailers' own brands. The food manufacturing sector plays a vital role in the supply chain being the link between producer and retailer and carrying the principal role in delivering quality, i.e. to deliver according to customer expectations in physical and financial terms. In the past, the food industry suffered because the linkage between producers and consumers was remote; more recently there has been much greater awareness of the interdependencies between each sector. The inputs to the manufacturing sector are now changing from a raw commodity to more clearly defined product status, there is a greater flow of information and development of relationships. The principal aims are to improve product safety, conformation and time scheduling.

Manufacturers now have increasingly closer working relationships with their customers, the multiple retailers, who demand precise quality standards, delivery schedules and consistency. Similarly, manufacturers now make increasing demands upon their suppliers (farmers and primary processors) for quality assured and cheaper produce.

The structure of the industry is dynamic, featuring mergers, take-overs and closures in response to pressure from retailers and the economies of scale and scope that can be achieved. The major manufacturers, such as Unilever, Associated British Foods, Cadbury, Nestlé, Northern Foods and Geest, have acquired smaller companies, while others, particularly in the declining canning and frozen food industries, have disappeared. This changing structure has implications for managers in the food supply chain, including the need to acquire the wide range of skills required to manage large workforces and the longer distances over which primary produce (especially perishable and sensitive goods) has to be transported to processing centres.

The five principal firms in the UK food retail sector are household names generally held in high esteem, and are a significant force in economic activity. Together, they enjoy more than 50% of the total market sales, and in 2002 employed 1.1 million or approximately 40% of all employees in retailing. In 2000, food and drink retailing combined accounted for 2.1% of UK gross domestic product. The retail sector is also dominated by the supermarket format and several strategies have contributed to its rise. Initially, the novelty factor and the freedom of individual selection at point of sale coupled with competitive pricing drove expansion; cost and convenience were principal strategies for supermarket growth. These worked well, but were less successful for fresh perishable goods, such as meat and vegetables which continued to be sold largely through butchers and greengrocers through to the 1980s. Another successful strategy was the development of larger stores on the edges of towns or on principal road networks of cities. The top UK food multiple retailers, who were simultaneously developing distribution networks, were quick to perceive the impact of increasing car ownership and the greater home food storage capacity offered by refrigerators and freezers, together with consumer preference for the one-stop weekly shop, and so grew to their current dominant position.

In the late 1990s, specialised retailers selling only organic produce emerged, and the food multiple retailers capitalised on this new food preference by developing separate organic produce sections in their stores.

Apart from the mainstream retail market, alternative sellers' markets are developing. The downward pressure on UK farm incomes resulting from the abundance of supply, the declining level of support under the Common Agriculture Policy (CAP), and the strength of the pound sterling that makes imports cheaper encourage the development of more direct sales through farmers' markets. Their impact on the retail market in volume terms is insignificant; however, they provide a valuable extra source of income for a minority of producers and a niche market. Their small presence is likely to increase. In contrast to French, German and Dutch agricultural producers who have a long tradition in co-operative processing and marketing, UK producers have taken less interest in direct sales, or in production of value-added products.

In addition, electronic shopping has experienced strong growth over the past few years due to the increasing use of the Internet (Ody, 1998). The top UK food multiple retailers offer it to their customers, although for the moment, the majority of their sales stem from traditional store retailing. The major challenge, however, is home delivery where food retailers deal with perishable products and products needing to be kept at various temperatures; a major concern is the risk of theft of the delivered products (McKinnon & Tallam, 2002).

All the foregoing has resulted in increasing retail power and subsequently in the erosion of the wholesaler's position in the food supply chain (McGoldrick, 2002). Traditionally, food wholesalers were responsible for purchasing fresh produce and manufacturers' products, and distributing them to retailers. This system has almost entirely been replaced by manufacturers distributing directly to retailers' warehouses (regional distribution centres). The next step is product distribution to retail stores for which retailers can use their own transportation and/or make use of third party logistics firms (Bourlakis, 1998). Logistics firms have experienced multinational growth during the past two decades following the deregulation of the transport industry, initially at national level and more recently at European Union level; they provide a range of traditional logistics services such as transportation and warehousing together with more advanced services such as inventory planning and information technology software development.

Overall, the changes taking place in the food supply chain, have affected manufacturers and primary producers who are still struggling to come to terms with them. Primary producers have increased their technical efficiency and receive support through the Common Agricultural Policy (CAP); they have produced more than most demand and, given the characteristics of the markets, their product prices have been driven down. Manufacturers have had to address the multiple retailers' strategy of developing their own brands that, through a combination of price and in-store promotion, have become increasingly important; unavoidably, the consequent loss of manufacturer's identity has impacted on the smaller, regional and local manufacturers.

Conclusions

This chapter has introduced the UK food supply chain and outlined its structure and contribution to the economy in terms of sales, value added and employment. Relationships within the chain are complex, have changed and continue to develop. Traditional structures and relationships are increasingly being replaced by larger scale, programmed, information-based sharing relationships which are determined by retailers' and manufacturers' standards.

The food market is, however, saturated with prospects of low growth. UK primary producers and manufacturers face increased competition from imports and are put under pressure to reduce prices by a highly concentrated retail sector. Economic analysis confirms that there is pressure on the food supply chain because price and income elasticity of demand for various food products are low or negative.

In the following chapters, the challenges faced by supply chain managers will be presented with a view to developing a better understanding of that continuously evolving discipline.

Study questions

1. Discuss the factors that influence modern food supply chains.
2. Analyse the importance of each chain member in terms of value added and employment.
3. Discuss the importance of price and income elasticity of demand for understanding food consumer behaviour.

References

Anon. (1999) Collaborating with electronic business. *Retail Solutions*, November 1999, pp. 20–21.

Bourlakis, M. (1998) Transaction costs, internationalisation and logistics: the case of European food retailing. *International Journal of Logistics: Research and Applications* **1**(3), 251–264.

Bowersox, D.J. & Closs, D.J. (1996) *Logistical Management: The Integrated Supply Chain Process*. McGraw-Hill, New York.

Christopher, M. (1998) *Logistics and Supply Chain Management: Strategies for Reducing Cost and Improving Service*. Financial Times, Prentice Hall, London.

Christopher, M. (1999) New directions in logistics. In *Global Logistics and Distribution Planning*, 3rd edn. (Walters, D., ed.), pp. 27–38. Kogan Page, London.

DEFRA (2002) *Strategy for Sustainable Farming and Food*. Department of Environment, Food and Rural Affairs, London (www.defra.gov.uk).

Economist Intelligence Unit (1988) *EDI for Retailers*. EIU Publications, London.
Fiddis, C. (1997) *Manufacturer–Retailer Relationships in the Food and Drink Industry: Strategies and Tactics in the Battle for Power*. Financial Times Retail & Consumer Publishing, Pearson Professional, London.
Fynes, B. & Ennis, S. (1994) EDI in retailing: implementation and prospects in Ireland. *International Review of Retail, Distribution and Consumer Research* **4**(4), 411–426.
Harvey, J. (1996) The basis of the UK's strength in international logistics. *Jubilee Lecture*, p. 4. Department of Business Studies, University of Edinburgh, UK.
Larson, P.D. & Lusch, R.F. (1990/1) Quick response retail technology: integration and performance measurement. *International Review of Retail, Distribution & Consumer Research* **1**(1), 17–35.
Lechene, V. (1999) Income and price elasticities of demand for meat, meat products & fish. In *National Food Survey*. Ministry of Agriculture, Fisheries & Food (MAFF), The Stationery Office, London.
McGoldrick, P. (2002) *Retail Marketing*, 2nd edn. McGraw-Hill, London.
McKinnon, A. & Tallam, D. (2002) Securing the last mile: an assessment of the risk of theft in home delivery channels. In *Proceedings of Logistics Research Network Conference*, 4–6 September 2002 (Griffiths, J., Hewitt, F. & Ireland, P. eds.), pp. 193–198. Institute of Logistics and Transport, Birmingham, UK.
Ody, P. (1998) *Non-store Retailing: Exploiting Interactive Media and Electronic Commerce*. Financial Times Retail and Consumer Publishing, London.
Quarrie, J. & Hobbs, S. (1997) *Supply Chain Technology: Improving Retail Efficiency and Effectiveness*. Financial Times Retail and Consumer Publishing, London.

Chapter 2

The Food Consumer and the Supply Chain

D. Marshall

Objectives

(1) To introduce the reader to different types of consumer buying decision.
(2) To identify some of the main factors influencing this decision making process.
(3) To outline the main stages in the consumer decision making process and discuss their relevance to food choice and the supply chain.
(4) To consider some specific food choice models.
(5) To examine food related lifestyles as one means of segmenting the market.

Introduction

Consumers make choices about food every day; these include decisions about whether to eat (or not), what to eat, when to eat, where to eat and who to eat with. These choices reflect their culture, their social status, and impact upon their physiological and psychological well-being. At one level, this behaviour is fairly mundane and routine; at another it is challenging, exciting and anything but routine. Consider, for example, the impact of food scares such as BSE and *E. coli* 157:H7 on the beef and egg markets and the way in which consumer decisions influenced the fortunes of those two sectors, or the ongoing debate about genetically modified foods, or concerns about the level of pesticides in fruit and vegetables. Our individual purchase decisions may seem relatively unimportant at the macro level but collectively it is the consumer, or more specifically the housewife (Miller, 1995), who determines success or failure in the food market. Early calls for agriculture to adopt a marketing perspective recognised the importance of looking towards the end of the supply chain and were based on the premise that production was ultimately driven by consumer demand. Better returns were available to those who acknowledged the sovereignty of the consumers and attempted to meet their needs. Food marketing was seen as a win–win situation, improving returns to the food industry and at the same time satisfying the various needs of different consumer groups.

The logic is sound, the rationale makes sense, but unfortunately too often it seems the consumer is simply at the end of a supply chain driven by a production ethos rather than one that gives due consideration to the end user. More often that not, only lip service is paid to the consumer. This may seem somewhat strange given the resounding appeal of the marketing concept, and the 'win–win' proposition. However, setting aside the inherent assumptions regarding consumer access to information, ability to process information, and adequate choice, one possible explanation for the limited focus on the consumer is that most of the participants in the food supply chain are not actually interested in the consumer! This is not simply a case of myopia, but reflects the fact that many organisations are dealing with customers, not final consumers, who have, arguably, different demands to the end user. The ability of the food supply chain to deliver what the end consumer wants depends on the degree of integration across the supply chain, the nature and direction of information flows, and the availability of near market information on what consumers actually want. Despite the recognition that consumers and technology represent twin trends driving much of the change within the food industry, many of the developments in recent years have focused on building relationships with other channel members. As one author commented (Hughes 1994), 'without the dynamics of consumer behaviour, and the new technical options, there would be little impetus to establish alliances or partnerships in the food industry'.

This chapter will focus on the end user, the consumer, the people who eat food purchased from supermarkets, corner stores, garage forecourts, restaurants, take-aways, and even on-line. The main aim of the chapter is to introduce some of the basic concepts underlying consumer decision-making and to consider how these ideas can be applied to the area of food choice and the implications for the supply chain. Readers are referred to the bibliography for a more detailed discussion of specific ideas and models.

Food consumption

The challenge for the food supply chain is to satisfy and meet consumer needs, wants and even their desires. To do this, it is essential to know what people are buying, to understand how they buy, and to explore the underlying reasons for their selection. The first question will not be addressed in any great detail in this chapter. There is a variety of government and commercial research describing what UK consumers eat: an example is the National Food Survey (2000) sponsored by the Department for the Environment, Food and Rural Affairs (DEFRA). Taking a snapshot of purchase behaviour in 2000, UK consumers spent £56bn on food (excluding alcoholic drinks). Expenditure on household food, soft drinks and confectionery (excluding alcoholic drinks) averaged £16.16 per person per week, an estimated 9.5% of consumers' total expenditure.

Table 2.1 shows consumption and expenditure for the main food groups in the UK between 1990 and 2000. The national average expenditure on food and drink (excluding alcoholic drinks) consumed outside the home was £7.36 per person,

Table 2.1 Consumption and expenditure for main food groups per person per week (National Food Survey, 2000).

	Consumption (g)[b]			Expenditure (pence)		
	1990	1999	2000	1990	1999	2000
Milk and cream[a] (ml or eq ml)	2169	2007	2081	127.7	131.8	136.2
Cheese	113	104	110	41.4	52.3	54.5
Meat and meat products	968	912	966	337.4	379.9	403.8
Fish	144	144	143	66.6	80.8	80.2
Eggs (number)	2.20	1.68	1.75	19.5	17.0	17.6
Fats and oils	255	186	186	36.1	34.8	35.9
Sugar and preserves	219	141	139	18.1	14.6	13.9
Vegetable and vegetable products[a]	2261	1966	1986	169.8	233.9	228.9
Fruit and fruit products	895	1063	1120	97.1	133.9	137.1
Cereal products	1470	1464	1508	203.9	271.1	279.0
Beverages	70	56	58	43.2	42.3	42.7
Miscellaneous	N/A	N/A	N/A	516	82.3	89.8
Total food (UK)	N/A	N/A	N/A	£12.12	£14.75	£15.20
Soft drinks (ml)[c]	N/A	1426	1523	N/A	55.7	59.2
Alcoholic drinks (ml)	N/A	385	435	N/A	130.0	149.7
Confectionery	N/A	53	64	N/A	28.2	55.6
Total all food and drink (UK)	N/A	N/A	N/A	N/A	£16.89	£17.64
Total all food (UK)	N/A	N/A	N/A	N/A	£14.75	£15.20
Total all food and drink (UK)	N/A	N/A	N/A	N/A	£16.89	£17.64

[a] 'Milk and cream' includes yoghurt, fromage frais and dairy desserts and 'Vegetable and vegetable products' includes potatoes unless otherwise stated.
[b] Except where otherwise stated.
[c] Converted to unconcentrated.
N/A: not available.

an increase of 13% in real terms since 1994. On average, three meals were eaten outside the home each week (National Food Survey, 2000). Table 2.1 shows consumption and expenditure across the main food categories and the proportionate consumption across each of the main food groups. Faced with derived demand and dependent on the end user, this type of information allows supply chain members to monitor general demand, identify trends and changes over time, measure consumer responsiveness to price, and get a general picture of what is happening in terms of consumption (measured here in terms of household purchases).

Commercial research is usually more selective in describing consumption and purchase behaviour of product categories and individual brands. This ranges from gathering usage or attitudinal data, through to consumer and sensory panels. Examples include the Taylor Nelson Sofres Family Food Panel which measures household consumption based on panel data from 4000 UK households – for more examples see Gofton (1998) and Lesser *et al.* (1986). Unfortunately, much of this commercial research is proprietary and few people outside the organisation concerned see it. In this respect, it is difficult to get an accurate picture of the extent to which supply chain companies are spending money researching the end user. The sceptics might say that the high failure rate of new food products suggests that we still have a lot to learn about the consumer. As well as asking what consumers are buying we need to discover why. Taking the case of beef and veal, Table 2.1 shows that consumption (defined here as household purchases) has returned to pre-BSE levels, but we do not know whether this is because UK consumers are more confident about their beef supplies, are convinced by reassurances from the government and suppliers that it is safe to eat, or have simply forgotten about this food scare and resumed their old habits.

Different types of buying decision

In looking at how consumers buy, Assel (1987) proposes four main types of buying decision based on the level of consumer involvement and the differences across brands (Box 2.1).

Much of the discussion on consumer decision-making has centred on complex buying behaviour where purchasers develop beliefs and attitudes about a product before making a considered choice. These types of decisions tend to be associated with expensive, risky, self-expressive and infrequently purchased goods. The other type of high involvement decision is dissonance-reducing buyer behaviour, where a consumer is highly involved in a product category but sees little difference between the brands on offer. The purchaser shops around looking for differences in quality or price. Alternatively, consumers may simply buy what is convenient, but after purchasing the product experience some psychological discomfort or 'dissonance' following exposure to new information or brands. Consequently, they may alter their beliefs or seek new information to support their purchase. Low involvement buying decisions include variety-seeking behaviour. In this case, an array of brand differences encourages consumers to switch their allegiance and consequently there is often little brand loyalty in the category. The motivation is the desire for variety rather than any lack of satisfaction with existing products. Finally, habitual buying behaviour reflects low involvement purchases in a situation where there are few distinguishable significant differences between brands. This is often the case with frequently purchased low cost products.

Box 2.1 Main types of consumer buying decision (adapted from Assel, 1987).

	High involvement	Low involvement
Significant differences between brands	Complex buying behaviour	Variety-seeking buying behaviour
Few differences between brands	Dissonance-reducing buying behaviour	Habitual buying behaviour

The questions are, where does food fit into all of this and what are the implications for the supply chain? Food is often seen as a low involvement product. For most consumers it is not expensive, accounting for less than 10% of total household expenditure. It is frequently purchased and, for the most part, is safe to eat, although as we shall see, food safety has become an important issue for consumers, with ramifications back through the supply chain. Most of our food purchases may be habitual – think about your weekly shopping trip and the time you spend in the supermarket. Similarly, variety seeking is recognised as a key element in food choice (Rozin & Markwith, 1991); this may arise from boredom, attribute satiation, or curiosity (Van Trijp, 1995), and marketing encouraging us to switch our allegiance and try new products. Conversely, there is some very interesting work looking at 'neophobics' who avoid unfamiliar and novel foods (Pliner & Hobden, 1992). In the case of low involvement products, easy access, product availability and avoiding out-of-stock situations are key concerns for the supply chain. Communication focuses on sales promotions to encourage trial, switching or to shift stock, while advertising campaigns strive to make the products more 'involving'. But while many would categorise food as a low involvement product, it can be highly involving for certain consumers, e.g. the gourmet, the 'foodie', the 'dieter', the recuperating and the bulimic, or certain situations, e.g. when entertaining, dining out at a restaurant or celebrating a special occasion. Even dissonance-reducing behaviour can be observed as consumers respond to adverse health messages about eating certain foods, try to make sense of food scares, and untangle conflicting messages from commercial and public organisations about what and how to eat. Under high involvement, consumers are more likely to engage in detailed information search, and to evaluate the products and the suppliers before they try; for example, by reading labels for information about ingredients, country of origin, and so on. We can see the problems and challenges of trying to categorise all foods within this general framework. The key is to try to understand how consumers regard your products across the entirety of their experiences. This requires us to examine how consumers become aware of our products, how they acquire them, how they prepare food, eat food, and even dispose of the food (MacMillian & McGrath, 1997; Marshall, 1995). This might involve us thinking introspectively, asking consumers about their behaviour retrospectively, prospectively, or even prescriptively (Kotler, 2003).

Factors influencing consumer choice

In modelling consumer behaviour we try to take account of the decision making process and consider how environmental and marketing stimuli impact upon this (Figure 2.1). Political, economic, social and technological factors clearly impact upon consumer decision-making; for example, the increase in affluence, the growth in single person households, changes in family size and the greater number of women in paid employment all have an impact on the type of food we buy. Consumer choice is shaped by cultural, social and personal characteristics, some of which are considered below.

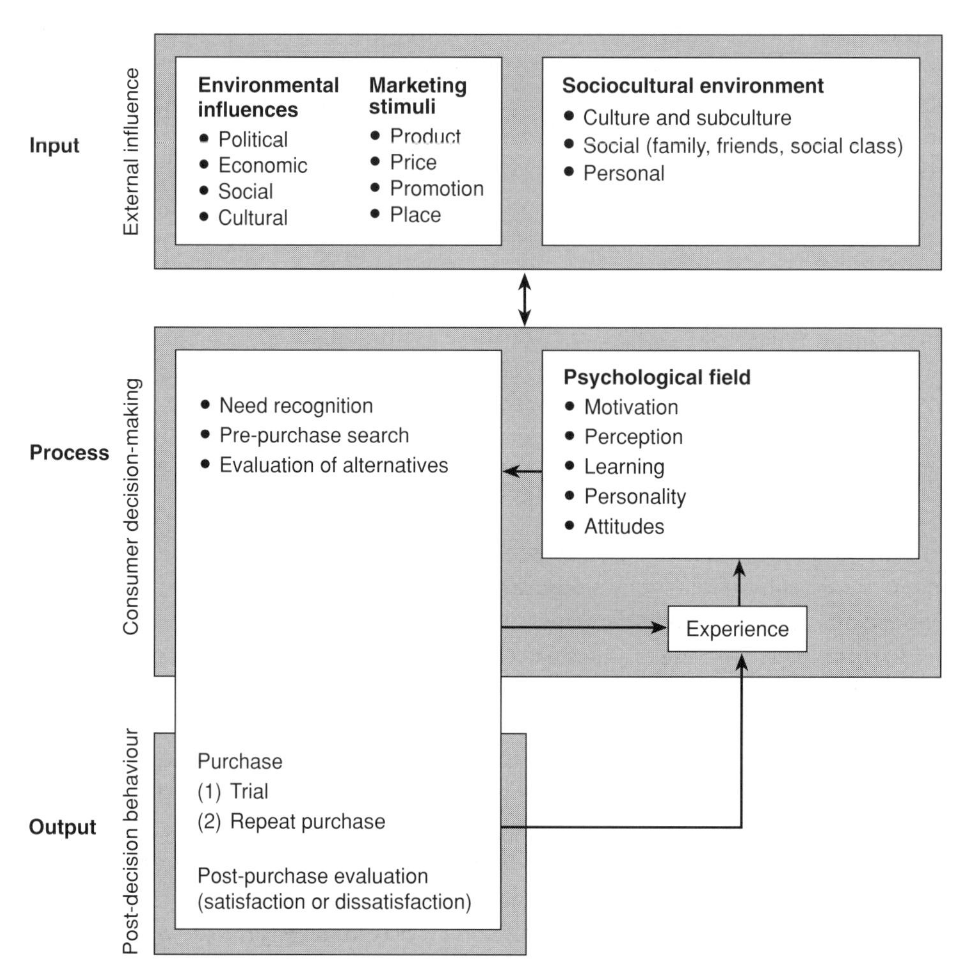

Figure 2.1 Basic model of consumer decision making (adapted from Shiffman & Kanuk, 2000).

Cultural influences include general mores, taboos and rituals that predetermined what is deemed appropriate and culturally acceptable (Falk, 1994). Hindus do not eat beef, Muslims reject pork, and few UK citizens would contemplate eating horsemeat or, worse, their pets. What we eat is governed by ritual; for example, each culture has prescribed patterns of eating and combining foods into recognisable dishes and sequences that they identify as meals. In the UK, this typically takes the form of meat, potatoes and vegetables, but there is evidence of more plurality in our meals and some redefinition of the idea of what constitutes a 'proper meal' in certain households (Marshall, 2000, 2003). Eating for many is essentially a social activity: the example of meals is a case in point and who we eat with can have a major impact on what is chosen (Makela, 2000). Sobal (2000) identifies the three main dimensions of sociability as facilitation (how people's eating is influenced by others), interaction (the social interchanges that occur at meals) and commensality (how eating partners are selected and excluded). This social context includes the company of family, friends, acquaintances and work colleagues, even strangers. In the family/household unit, individuals often take on different responsibilities and buying roles can be identified as follows (adapted from Kotler, 2003):

(1) *Initiator*: The person who first suggests the idea of buying the food product or service.
(2) *Influencer*: The person whose view or advice influences the decision.
(3) *Decider*: The person who decides on any component of a buying decision and whether to buy, what food to buy, how to buy or where to buy.
(4) *Buyer*: The person who makes the actual food purchase.
(5) *User*: The person who consumes or uses the food product or service.
(6) *Gatekeeper*: The person who regulates exposure to information and products intended for the family/household.

The nature of decision making within the household depends on the type of family unit, but it seems that in the case of food, women continue to take the bulk of responsibility for food shopping (Davis & Rigeaux, 1974).

They often act as 'gatekeepers', introducing household members to new products and making choices that accommodate the wants and preferences of their partners and family. However, in our own research among young married couples we found that men still exert significant control over what is purchased, which raises some questions about the relevance of the female 'gatekeeper' role in these households. It is reasonable to assume that they act as conduits for the satisfaction of their partner's wants and, at least in these young couples, continue to accommodate their partner's wishes in a variety of implicit and explicit ways (Marshall & Anderson, 2000). There is evidence to suggest that more men are going food shopping, either as single person households or as members of a family or household unit, and it appears that this trend is set to continue, particularly where the partner's main occupation is not that of homemaker (Maret & Findlay, 1984; Ekström, 1991; Warde & Hetherington, 1994; Sullivan, 1997; Dholakia *et al.*, 1995; Marshall & Anderson, 2000). What is less clear is the precise role they play in this task. This question of whether we are dealing with individual or collective decisions is important, and we know relatively little about what happens within the household in terms of food distribution and domestic division of labour. Much of the focus has been on buying food, with much less attention directed towards the distribution of food within the household unit, meal patterns and food use, or the extent of individualised eating and independence within the domestic unit. The research needs to move beyond the point of purchase and consider how products are being used in the household and by whom.

Finally, much of the focus on consumer behaviour has centred on the individual consumer, looking at motives, attitudes, how consumers learn, and trying to equate personality and lifestyle with consumption. Measuring consumer attitudes remains an important part of understanding consumer choice. Shepherd (1989, 2001) has been looking at the relationship between consumer attitudes and food choice using structured attitude models such as the Ajzen and Fishbein theory of reasoned action (TRA) or the Ajzen theory of planned behaviour (TPB). Both attempt to model the relationship between attitudes and behaviour, and argue that intention to perform behaviour is the best predictor of behaviour. This measure comprises a person's attitude, the social pressure to conform, and in the case of the TPB, an element of perceived control. While TPB generally shows good behaviour prediction and can identify important dimensions of driving food choice, the relationship between simple taste and dietary intake is not particularly strong, and other factors need to be taken into account (Shepherd, 2001).

Stages in the buying decision process

Marketers are interested in who makes the buying decision, the type of decision and the various stages in that process. Most 'hierarchy response models' break the buying decision down into input, process and output variables, with a series of five interrelated stages: problem recognition, information search, evaluation of alternatives, purchase decision and post-purchase behaviour (Nicosia, 1966; Howard & Sheth, 1969; Engel *et al.*, 1991). These models assume that highly involved buyers go through a cognitive, affective, behavioural response to highly differentiated products. The early stages of problem recognition begin when a buyer recognises a need: in the case of food this may be triggered by a basic physiological drive such as hunger or thirst; it may be also triggered by an external stimulus, for example, seeing other people eating food, or eating because of a prescribed social or cultural practice such as eating at particular times of the day, or consuming certain foods on special or celebratory occasions. The search for information depends on the problem or need stage, but essentially leads to a heightened awareness and may induce an internal or external search for information. In the former case, individuals will use existing information in their memory to recall certain food products or brands that they have used in the past and may choose on the basis of experiential sources. If this information is insufficient, for example a reasonable length of time has elapsed since they last purchased a particular item, consumers may seek external sources of information from personal sources, asking friends or family or other associates for information and advice. They may consult commercial sources, such as advertising, salespeople, shop assistants, merchandising display and packaging information, or websites. Other sources of information include public sources such as the mass media, and consumer associations. In one survey, the most trusted source of off-label information for UK consumers was store leaflets (41% cited this as their first mention), followed by newspapers (15%) and TV/radio news (14%) (IGD, 2002). While a large amount of information, such as general media coverage, is available in the public domain and through commercial sources, most trust is placed on personal sources. The advantages here are that the communication is interactive and two-way, and in the case of word-of-mouth recommendations from friends, peers and opinion leaders the information is (usually) independent. Food manufacturers are the least trusted source of information; the most trusted are GPs/practice nurses (IGD, 2002).

Through past experience and information search, consumers build up a level of awareness about what products and brands are available in the market (awareness set), know which ones meet their buying criteria (consideration set), decide on the main brands that meet their needs (choice set) and use this to make a final decision on what to purchase (Nedungadi, 1990). One of the key issues for those in the supply chain is to determine how many of the target market know about their brands and include them in their consideration and choice sets. Recent research has shown that product success often depends on getting the balance right between matching the product qualities of the leading brands in a category and creating a significant point of differentiation, but not enough to exclude the product from the consideration set (Garber *et al.*, 2003). Moreover, what is included in the consideration set may vary according to the situation (Marshall, 2003). This might explain the proliferation of me-too products in a marketplace where retailers are developing and expanding own-label brands to rival or surpass category leaders and then competing on price.

Having acquired information about products and brands, consumers make judgements about the products in terms of how well they satisfy their needs, how the product attributes match their requirements and what benefits they can deliver. Typically this involves identifying key attributes. In the case of food, much product testing focuses on sensory systems involved in flavour perception, notably taste (gustation), smell (olfaction) and common chemical sense (trigeminal). Characteristically, taste is seen as comprising sweet, sour, salty and bitter qualities, with 'umami' included as the fifth basic element in Japanese culture. Sensory research essentially focuses on examining the key sensory elements that affect consumer liking and determining the key sensory elements driving consumer choice. Our ability to select from a range of foods may play an important role in ensuring dietary variety (Rogers & Mela, 1992). As Rozin (1989) has shown, many of our food preferences are learned, and this plays an important role in food choice. Aversions to certain foods may result from negative experiences affecting our liking or disliking them and determine what is excluded and included in our consideration sets. Product evaluation goes beyond sensory aspects, and brand attributes might include marketing variables such as price, value for money and convenience (Garber *et al.*, 2003). Research by the UK Food Standards Agency (2001) found consumers rated price (46% of adults) as the most important (unprompted) factor that influenced their food buying, with taste (18% of adults) coming second. Other factors mentioned by respondents included quality, family health, production methods, appearance, freshness, health/nutritional value, brand name and food safety. When prompted, respondents rated family health (98% of adults), taste (97%), food safety concerns (93%), conditions in which animals were raised (88%), environmental concerns (88%), price (78%) and appearance (78%) as very/quite important factors.

Increasingly, the point of differentiation lies in aesthetics or what has been called the 'sugar coating' on commodities (Welsch, 1996). What is more, unlike sensory differences, consumers can readily distinguish between products on these non-sensory criteria. It is the 'sugar coating' and the symbolic aspects of consumption that are driving choice. To reiterate the point, consumers do not make choices on the basis of (sensory) taste alone, although somewhat ironically taste has become the new point of differentiation for some brands, as they compete on taste as the new unique selling point. Recent examples include Sprite commercials (e.g. Image is nothing, Obey your thirst). These attributes contribute to the brand image, but much depends on the consumer's experience with the product and the effect of selective perception, distortion and retention of information. Rather than some universal significance, brands have multiple meanings and carry different significance for different people (Elliot & Wattansuwan, 1998).

The purchase stage requires consumers to make decisions about the brand, the seller, how much to buy and when to buy, as well as how to pay for the product. The post-purchase stage depends very much on the consumer's experience with the product; if it fails to meet expectations dissatisfaction may arise which, in the case of food, may relate to taste, flavour, texture, portion size, preparation and a whole host of other factors. Disconfirmed expectations are particularly problematic and raise a number of questions about the extent to which consumer expectations are raised by marketing activities (Cardello, 1994). Disgruntled customers are more likely to spread negative information by word of mouth, emphasising the need to consider what happens after the point of sale. Freephone numbers, customer helplines and feedback surveys all help to monitor consumers' experience with products; dealing effectively with complaints is essential to retain consumer business. A major challenge is to be able to track the complaint back through the supply chain and deal with it effectively.

These general consumer decision models have been criticised for lacking specificity, not being tested, or not presenting testable hypotheses, and for the assumptions made about the rational nature of the decision-making process (Tuck, 1976; Foxall, 1991). Consumers are, for example, assumed to go through these stages in an ordered fashion before reaching a final decision, but the proponents of the low involvement models suggest that in certain cases consumers may act on the basis of emotion and evaluate the products after the purchase. In the case of food choice, one might assume that the nature of the process makes these complex decision models much less relevant. However, as stated earlier, there are situations where they may be deemed appropriate; food scares such as BSE have created a heightened awareness of the risks associated with food choice and for many consumers forced a much more involved relationship with food. It has been argued that we may be unwise in attributing choice processes to consumers based on their use of information before purchase, as illustrated by their limited willingness or ability to comprehend nutritional information in making food purchase decisions (Jacoby, 1977). Many of our food purchases involve little deliberation, and may simply arise from necessity, childhood preferences, social conformity and word-of-mouth recommendation, or random events (Olshavsky & Granbois, 1979). However, in trying to apply these general models of consumer decision making, one has to remember

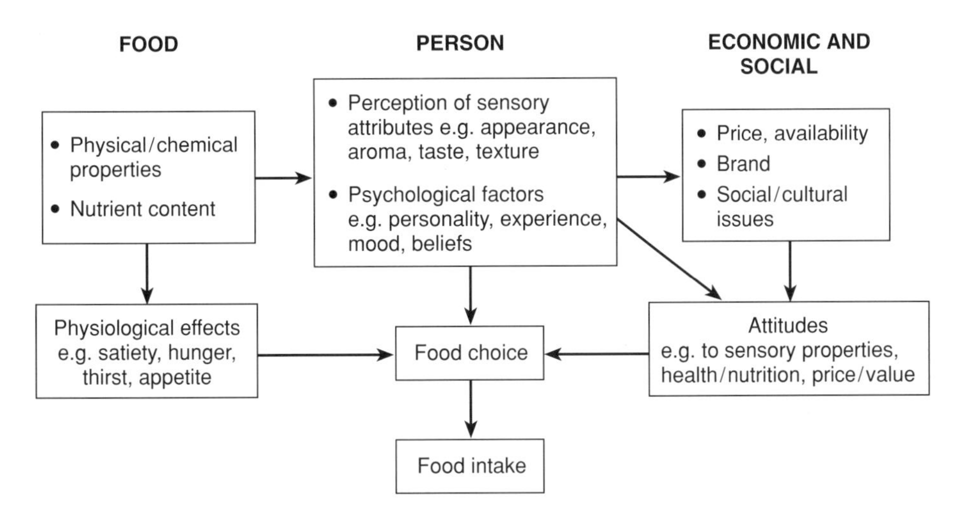

Figure 2.2 Factors affecting food choice and intake (Source: Shepherd, 1989, 2001).

that food is a somewhat unique commodity, in the sense that it is actually ingested and literally becomes part of you (Falk, 1994). This raises the question about whether we should regard it as a low involvement commodity. However, the high frequency of most food purchases means that after initial exposure there is little need to go through an extensive decision-making process and, for the most part, consumers have to trust those who supply the products from farm gate to the plate. Ultimately, it depends on how consumers regard their level of involvement with food, and the degree of consumer confidence depends on how much trust consumers place in the supply chain.

A number of specific food choice models have focused on cognitive and motivational elements (Krondal & Lau, 1982; Michela & Contento, 1986), whilst some have included social factors (Worsley *et al.*, 1983). More comprehensive models (Figure 2.2) consider the relationship between personal characteristics, environmental features and the food product (Parraga, 1980; Shepherd, 1989). These models are very product focused and centre on the relationship between taste and flavour in relation to these other influences. Furst *et al.* (1996) looked at how consumers viewed their own food related behaviours and constructed a conceptual model that grouped factors involved in food choice into three main components: (a) life course, (b) influences and (c) personal system; the inter-relationships between these determine what is chosen. Life course includes social, cultural and physical environments to which the consumer is exposed and also personal roles. This generates influences that inform and shape personal systems. It is the life course that gives rise to the influences that emerge in a food choice situation and the ways in which individuals execute that choice under different social and physical settings.

Consumer segments: the who?

Marketing not only advocates addressing consumer needs, but also demands that we clearly identify which consumers we aim to please. Market segmentation is about identifying groups of consumers with similar wants and needs, and then catering specifically to those needs. The basis of this segmentation may be demographic (age, gender, social class, occupation, income, ethnic group), geographic, geo-demographic, psychographic/lifestyle, or behavioural (brand loyalty, usage situation, benefits) (Solomon *et al.*, 2002). Each aims to identify distinct groups of buyers that become the target for marketing activities. This is where the consumer/customer distinction is important as most of the channel members have to cater for their customers' needs rather than concern themselves with the end consumer. Most of these segmentation bases are self explanatory, but psychographic or lifestyle segmentation looks at differences in terms of consumer activities, interests and opinions, and reflects differences in consumer attitudes or values. Lifestyles are patterns of consumption that reflect how consumers spend their time and money. National or linguistic borders are significant in defining food consumption patterns, and one study has identified twelve general food cultures in Europe (Askegaard & Masden, 1998). Another application is the food related lifestyle (FRL) concept: it is a person related construct that complements the more established measures such as attitudes and beliefs, or personality traits such as variety seeking or neophobia, and can help explain differences in patterns of choice across a range of products (Figure 2.3). The model is an intermediate cognitive construct located between more abstract life values and concrete product/brand perceptions or

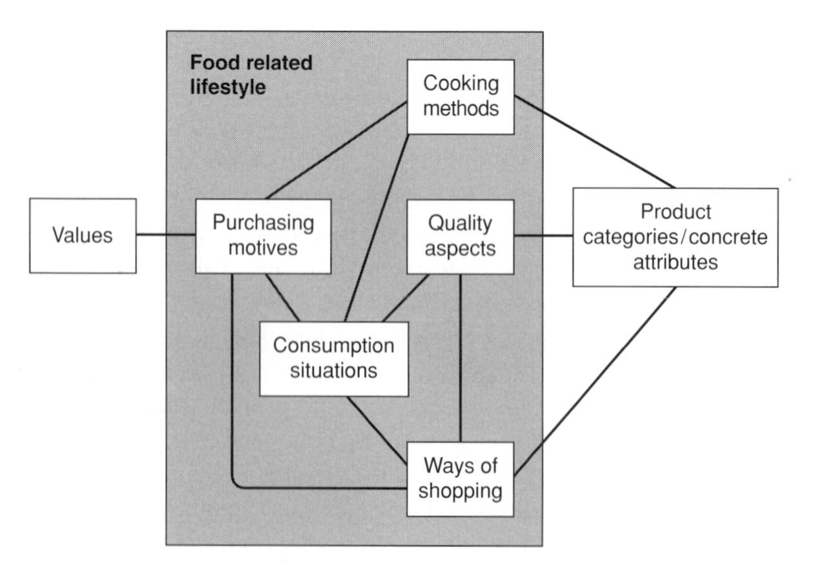

Figure 2.3 Cognitive structure model for food related lifestyle (adapted from Grunert *et al.*, 1995).

attitudes, and comprises various dimensions that include ways of shopping, quality aspects, cooking methods, consumption situations and purchasing motives. It has been applied in representative consumer surveys in Denmark, France, Germany and the UK.

A number of common segments have been found across the European countries, and the UK study identified the following five basic segments, with figures in parentheses reflecting the proportion of the UK population in each segment (adapted from Brunso *et al.*, 2002; Grunert *et al.*, 1995):

(1) *The uninvolved food consumer (9%)*: For these consumers, food is not a central element in their lives. Consequently, their purchase motives for food are weak, and their interest in food quality is limited mostly to the convenience aspect. They are also uninterested in most aspects of shopping, do not use speciality shops, and do not read product information, limiting their exposure to and processing of food quality cues. Even their interest in price is limited. Their interest in cooking is limited as well, they do not plan meals and snack a lot.
(2) *The careless food consumer (27%)*: These consumers are in many aspects like the uninvolved food consumer, meaning that food is not very important to them, and their interest in food quality, with the exception of convenience, is correspondingly low. The main difference is that these consumers are interested in novelty: they like new products and tend to buy them spontaneously, as long as they do not require a great effort in the kitchen or new cooking skills.
(3) *The conservative food consumer (19%)*: For these consumers, the maintenance of security and stability in life by following traditional meal patterns is a major purchase motive. They have major interests in taste and health of food products, but they are not particularly interested in convenience, since their meal preparation is traditional and regarded as part of a woman's task.
(4) *The rational food consumer (33%)*: These consumers process a lot of information when shopping; they look at product information and prices, and they use shopping lists to plan their purchases. They are interested in all aspects of food quality. Self-fulfilment, recognition and security are major purchase motives for food for these consumers. Also their meal preparation is characterised by planning and reason.
(5) *The adventurous food consumer (12%)*: These consumers have a somewhat above average interest in most quality aspects, but their profile is most pronounced with regard to meal preparation. They are very interested in cooking, look for new recipes and new ways of cooking, involve the whole family in the cooking process, are not interested in convenience and reject the notion that cooking is the woman's task. They want quality and demand good taste in food products. Self-fulfilment in food is an important purchase motive. Food and food products are an important element in these consumers' lives. Cooking is a creative and social process for the whole family.

The segmentation results from the UK show how the different segments relate to food and suggest that a variety of decision-making processes co-exist, depending on the consumer's lifestyle and how he or she relates to food provisioning. While the FRL concept does not explicitly discuss the nature of the decision making process, it seems reasonable to assume that certain segments, e.g. the conservative and adventurous segments, are likely to be much more highly involved in the decision making process compared to the uninvolved and careless. Over half of the UK population is more or less uninterested in cooking or shopping! The FRL concept has a number of applications in the area of new product development. In one application, a fish processing company targeted adventurous and rational consumers for a range of new fish-based meals based on FRL segment profiles for the UK and French markets. This research illustrates the possibility of defining cross-cultural segments on the basis of psychological rather than behavioural variables. Similar purchase behaviours may reflect very different motivations and this adds a further perspective to consumer food choice.

Limits on choice

Of course, consumer choice depends on what is available, and this is influenced by a range of physical, economic and political elements (Fieldhouse, 1996). Food availability reflects decisions made by both private and public sector organisations across the supply chain (retailers, manufacturers, etc.). While policy emphasis has shifted towards improving safety and quality amidst heightened consumer awareness, there is little evidence that consumers have had much impact on agricultural policy, despite calls for greater consumer voice (Flynn *et al.*, 1998). In the UK, the establishment of the Food Standards Agency (FSA) in April 2000 was one attempt to address domestic consumer issues independently of producer interests: the agency aims to prioritise public interest and 'put the consumer first' by providing advice and information to consumers and government on matters such as food safety across the food chain, nutrition and diet, accurate and meaningful labelling, and ensuring that food law is effective and enforced. The FSA will complement and support the European Food Authority (Anon., 2000), as well as increasing public involvement in regulating issues to do with food safety. A recent Commission of the European Communities White Paper reflects this emphasis on safety, with calls for the establishment of an independent European Food Authority to guarantee a high level of food safety. It considers the need for community-wide food safety legislation, food safety controls and consumer information. It is difficult to determine the extent to which consumers have had an impact on food policy, and there is a need for more robust and comprehensive criteria to determine whether or not various mechanisms for public participation have been successful (Rowe *et al.*, 2001).

Consumers have become increasingly distanced from the farm gate and even more reliant on the food industry to supply them with their food. The incidence of food poisoning, BSE, *E. coli* 157, genetically modified foods, pesticide levels in fruit and vegetables, and exposés on the fast food industry have all heightened consumer awareness. These concerns are not only with safety and quality issues, but extend to labour conditions and working practices (Klein, 2000; Scheauser, 2000). When a book on fast food becomes a best seller you can be sure that consumers are taking notice. Consumers can make an impact through their buying behaviour and some are beginning to question how their food has been grown, reared, processed or packaged, or how it is served, stored or handled. Consumer concerns over the introduction of genetically modified ingredients played a key role in the decision to withdraw brands with genetically modified ingredients from supermarket shelves (Davis, 2001). Furthermore, the interest in farmers' markets, 'slow food', sustainability and organics is not main stream, but it may be the start of consumers seeking alternative sources of supply. With (arguably) little to distinguish between the quality, taste and price of competing brands, consumers are looking at how firms conduct their business between the (agricultural) fork and the plate. Consumers are beginning to rattle the chain and those organisations supplying these highly competitive markets will feel the repercussions.

Understanding food choice requires companies to adopt a broader perspective on consumer food choice. Food supply chains are built around individual products or brands, but consumers eat a wide variety of food. Supply chains ultimately converge at the point of consumption, when consumers make the final decision to eat. Take the case of meals, mentioned earlier. Most of us eat our food as part of a meal and, despite the trend in snacking, this remains a fundamental part of how we 'do food'. With the current interest in 'meal solutions' the boundaries between the retail and food service sectors have become blurred as food preparation moves out of the home and 'eating out' moves into the home. Surely this is a good time to look at what happens after purchase and how consumers use food products as part of their provisioning process. Moreover, the food buyer (acquisition) and the food consumer (use and disposal) are not necessarily the same person. It appears that most of the meal solution developments are being driven by the food service sector with endorsement by celebrity chefs rather than grounded in an understanding of what happens in the household.

Consumers are one of the main forces driving change in the food industry (Box 2.2), but the extent to which they do this depends on how well the industry

Box 2.2 Consumer forces driving partnerships and alliances in the European food industry (adapted from Hughes, 1994).

(1) Population and demographic change
(2) Incomes and levels of living
(3) Social change and family work patterns, and cultural attitudes to work
(4) Increasing consumer awareness of health, animal welfare, and environmental issues
(5) Growing ethnic diversity
(6) The pleasure principle

monitors and responds to the end consumer. But one question is, who is going to take the initiative on this? It could be argued that retailers are in the best position to understand consumer food choice, not simply because of their proximity to the consumer but also due to the fact that they have amassed vast amounts of data on consumer purchase behaviour (too much to know what to do with it!). Despite claims about addressing consumers' needs, and the proclamations about efficient consumer response (ECR), these programmes appear to do more for the retailers and their appointed suppliers than consumers. While there is some question over the extent to which consumers have any real sovereignty in a market where retailers exert considerable control over the supply chain and what is available (Marshall, 2001), they do have access to more independent information and can assess what companies are doing. And, if they don't like it, they can always shop elsewhere (Scansaroli & Szymanski, 2002). Purchase may be seen as the final stage in the supply chain, but for consumers it is just the start of their engagement with food that has to be stored, prepared, cooked, and often combined with other foods, before it is eaten and disposed of (Marshall, 1993). The retail stage is the main point of engagement with the supply chain for most consumers, but it is only the beginning of their involvement as food consumers. Many are beginning to look back along the supply chain and ask questions about the processes by which food makes it to their store, their refrigerator and their plate. Those further down the supply chain may choose to ignore what happens at the consumer's end of the chain and focus on building better relationships with their customers. However, the trend towards globalisation, further distancing consumers from production, growing concerns over health and safety, a lack of trust, ethical concerns and questions of sustainability, along with the search for authenticity, are all likely to have a major impact on the supply chain(s).

The future

So what of the future? The new consumer seeks authenticity, individuality, involvement, independence and wants to be well informed. The challenge is to give consumers the time and attention they demand and to build their trust (Lewis & Bridger, 2000). In Box 2.3 six (US) consumer trends are identified and the implications for retailers considered. While technology has had a major impact on business practice and consumer shopping, 'in many instances, the application of technology has occurred in the absence of a keen understanding of consumer trends and values. The results have been mixed at best and disappointing in many other instances' (Scansaroli & Szymanski 2002, p. 7).

Those involved at all stages in the supply chain, not only retailers, need to be aware of these trends and attempt to build closer relationships with consumers that genuinely work towards meeting their needs. In an increasingly diverse marketplace identifying, categorising and understanding 'the food consumer' is becoming evermore challenging. Amidst all this change, consumers need reassurance and some degree of continuity in their eating patterns and food habits.

Box 2.3 Consumer trends and new retail values (adapted from Scansaroli & Szymanski, 2002).

Consumer trends

(1) Household diversity
- wide variety of structure of households – nuclear family, same sex groups, singles, etc.
- variety of ethnic and cultural backgrounds

(2) Behaviour clues have changed
- individuality of lifestyle means young 'old' people, 'old' young people, house-husbands, etc.
- shorter attention span, more informed
- forget the assumptions

(3) Consumer power has shifted
- more empowered
- ability to assess fairness of transactions
- aware of opportunity cost of time – many views of what constitutes convenience
- expect to be rewarded

(4) Stimulation and sanctuary
- balance between these
- 'real' experiences
- desire for retreat, reflection, comfort and security

(5) Amorphous codes and spaces
- no norms – single parents, elder care, etc.
- wear multiple hats – roles not clearly specified
- redefinition of spaces – home as office, car as office, etc.

(6) Communities everywhere
- big variety with different aims – virtual and real
- multiple community membership – multiple influences on consumers
- consumer values communities more now

New retail values

(1) Consumer respect as central to success not tangential
- respect for individual lifestyles
- respect for consumer time
- respect for consumer space
- respect for consumer choices

(2) Soul of the customer, not superficial understanding
- extended role for customer relationship management
- deep understanding of consumer

(3) Customer enthusiasm, not customer satisfaction
- outcomes exceed expectations
- development of customer advocates
- enthusiasm results in innovation

(4) Customers customise, not retailers
- customers exercise power
- need to customise the experience
- more complex retail environments and management

(5) Community leadership, not participation
- leading causes not just supporting
- important when entering new markets
- customer communities

Acknowledgements

Thanks to Professor John Dawson, University of Edinburgh, and Joachim Scholderer, MAPP (Centre for Research on Customer Relations in the Food Sector, www.mapp.asb.dk), for their help in sourcing some of the material for this chapter.

Study questions

1. Using Assel's model, select a food product to represent each of the four types of buying decision. What difficulties do you encounter trying to categorise food products in this way?
2. Identify the major cultural, social and personal influences on what you eat.
3. List all the information sources you consulted over a recent food purchase. Can you characterise these (internal, external, etc.) and evaluate their relative impact on your final decision.
4. a. Try to identify current consumer trends or issues that are likely to have an impact on UK supply chain(s).
   b. How do these compare to the six (US) consumer trends identified by Scansaroli & Szymanski (2002)?
5. What did you eat yesterday? How and why did you choose those items and the place you ate them? Can you locate these decisions in the conceptual frameworks provided in this chapter?

References

Anon. (2000) FSA ready to rumble. *The Grocer* 8 April, p. 30.

Askegaard, S. & Madsen, T.K. (1998) The local and the global: exploring traits of homogeneity and heterogeneity in European food cultures. *International Business Review* **7**(6), 549–568.

Assel, H. (1987) *Consumer Behavior and Marketing Action*. Kent Publishing, Boston.

Brunso, K., Fjord, T.A. & Grunert, K.A. (2002) Consumer food choice and quality perception. *MAPP Working Paper Series*. Aarhus School of Business, Denmark.

Cardello, A. (1994) Consumer expectations and their role in food acceptance. In *Measurement of Food Preferences* (H.J.H. MacFie & D.M.H. Thomson, eds), pp. 253–297, Blackie, London.

Davis, H.L. & Rigaux, B.P. (1974) Perception of marital roles in decision process. *Journal of Consumer Research* **1** (June) 5–14.

Davis, S. (2001) Food choice in Europe – the consumer perspective. In *Food, People and Society: A European Perspective of Consumer's Food Choices* (L. Frewer, E. Risvik & H. Schifferstein, eds). Springer-Verlag, Berlin.

Dholakia, R.R., Pedersen, B. & Hickmet, N. (1995) Married males and shopping: are they sleeping partners? *International Journal of Retail and Distribution Management* **23**(3), 27–33.

Ekström, M. (1991) Class and gender in the kitchen. In *Palatable Worlds: Sociocultural Food Studies* (E.L. Fürst, R. Prättälä, M. Ekström, L. Holm & U. Kjaernes, eds). Solum Forlag, Oslo.

Elliot, R. & Wattanasuwan, K. (1998) Brands as symbolic resources for the construction of identity, *International Journal of Advertising* **17**(2), 131–144.

Engel, J.F., Blackwell, R.D. & Miniard, P.W. (1994) *Consumer Behaviour*, 8th edn, Dryden, Fort Worth.

Falk, P. (1994) *The Consuming Body*. Sage, London.

Family Expenditure Survey (2000) DEFRA, London. (http://www.defra.gov.uk/esg/work_htm/publications/cf/fes/fes.htm)

Fieldhouse, P. (1996). *Food and Nutrition: Customs and Culture*, 2nd edn. Chapman and Hall, London.

Flynn, A., Harrison, M. & Marsden, T. (1998) Regulation, rights, and the restructuring of food choices. In *The Nation's Diet: the Social Science of Food Choice* (A. Murcott, ed.), pp. 147–151. Longman, London.

Food Standards Agency (2001) *Food Concerns Omnibus Survey*, prepared for Food Standards Agency by COI Communications, 27th September 2001, located at http://www.talkfood.org.uk/data/consumer-research/index.html.

Foxall, G. (1991) Consumer behaviour. In *The Marketing Book*, 2nd edn (M. J. Baker, ed.), pp. 193–214. Butterworth-Heinemann, Oxford.

Furst, T., Connors, M., Bisogni, C.A., Sobal, J. & Falk, L.W. (1996) Food choice: a conceptual model of the process, *Appetite* **26**, 247–266.

Garber, L.L., Hyatt, E.M. and Starr, R.G. (2003) Measuring consumer response to food products. *Food Quality and Preference* **14**(1), 3–15.

Gofton, L.R.G. (1998) British market research data on food: a note on their use from academic study of food choice. In *The Nation's Diet: the Social Science of Food Choice* (A. Murcott, ed.), pp. 302–310. Longman, London.

Grunert, K.J., Brunso, L. & Soren Bisp (1995) Food related lifestyle in Great Britain, *MAPP Project Report No. 14*. Aarhus School of Business, Denmark.

Howard, J.A. & Sheth, J.N. (1969) *The Theory of Buyer Behaviour*. Wiley, New York.

Hughes, D. (ed.) (1994) *Breaking with Tradition: Building Partnerships and Alliances in the European Food Industry*, p. 6. Wye College Press, London.

IGD (2002) *Food Consumption: The One Stop Guide to the Food Consumer*, IGD Business Publications, Institute of Grocery Distribution, Watford.

Jacoby, J., Chestnut, R.W. & Silberman, W. (1977) Consumer use and comprehension of nutrition information, *Journal of Consumer Research* **4**, 119–128.

Klein, N. (2000) *No Logo*. Flamingo, London.

Kotler, P. (2003) *Marketing Management Analysis Planning and Control*, 11th edn, Prentice Hall, Englewood Cliffs, NJ.

Krondal, M. & Lau, D. (1982) Social determinants in human food selection. In *The Psychobiology of Human Food Selection* (L.M. Barker, ed.), pp. 139–151, AVI Publishing, Westport.

Lesser, D., Hughes, D. & Marshall, D. (1986) Researching the food consumer – techniques and practice in the UK and North America. In *The Food Consumer* (C. Ritson, L. Gofton & J. McKenzie, eds), Wiley, Chichester.

Lewis, D. & Bridger, D. (2001) *The Soul of the New Consumer: Authenticity: What We Buy and Why in the New Economy*. Nicholas Brearley, London.

MacMillian, I.C. & McGrath, R.M. (1997) Discovering new points of differentiation, *Harvard Business Review*, July/August, 133–145.

Makela, N. (2000) Cultural definitions of the meal. In *Dimensions of the Meal: The Science, Culture, Business and Art of Eating* (Meiselman, H.L., ed.), pp. 7–15. Aspen, Gaithersburg, MD.

Maret, E. & Findlay, B. (1984) The distribution of household labour among women in dual earner households, *Journal of Marriage and the Family* **46** 357–364.

Marshall, D.W. (1993) Appropriate meal occasions: understanding conventions and exploring situational influences on food choice, *International Review of Retail, Distribution & Consumer Research* **3**(3), 279–301.

Marshall, D.W. (1995) *Food Choice and the Consumer*, Chapman and Hall, London.

Marshall, D.W. (2000) British meals and food choice. In *Dimensions of the Meal: The Science, Culture, Business and Art of Eating* (H.L. Meiselman, ed.), pp. 202–220. Aspen, Gaithersburg, MD.

Marshall, D.W. (2001) Food availability and the European consumer. In *The European Consumer* (L. Frewer, E. Risvik & R. Schifferstein, eds), pp. 317–336. Springer Verlag, Berlin.

Marshall, D.W. (2003) Commentary on Garber *et al.* measuring consumer response to food products, *Food Quality and Preference* **14**(1), 17–21.

Marshall, D.W. & Anderson, A.S. (2000) Who's responsible for the food shopping, a study of young Scottish couples in their honeymoon period. *International Review of Retail, Distribution and Consumer Research* **10**(1), 59–72.

Michela, J. & Contento, I. (1986) Cognitive, motivational, social, and environmental factors on food choices, *Health Psychology* **5**, 209–230.

Miller, D. (1995) Consumption as the vanguard of history: a polemic by way of introduction, In *Acknowledging Consumption: A Review of New Studies* (D. Miller, ed.), pp. 1–57. Routledge, London.

National Food Survey (2000) Annual report on food expenditure, consumption and nutrient intakes. DEFRA, London. (http://www.defra.gov.uk/esg/Work_htm/publications/cf/nfs/nfs.htm)

Nedungadi, P. (1990) Recall and consumer consideration sets: influencing choice without altering brand evaluations, *Journal of Consumer Research* **17**, 263–276.

Nicosia, F.M. (1966) *Consumer Decision Processes*. Prentice-Hall, Englewood Cliffs, NJ.

Olshavsky, S.W. & Granbois, D.H. (1979) Consumer decision making – fact or fiction? *Journal of Consumer Research* **6**, 93–100.

Parraga, I.M. (1980) Determinants of food consumption, *Journal of the American Dietetic Association* **90**, 661–663.

Pliner, P. & Hobden, K. (1992) Development of a scale to measure the trait of food neophobia in humans, *Appetite* **19**, 1105–1120.

Rogers, P.J. & Mela, D.J. (1992) Biology and the senses: do you eat what you like or like what you eat? In *Your Food: Whose Choice?* (National Consumer Council, ed.), HMSO, London.

Rowe, G., Reynolds, C. & Frewer, L. (2001) Public participation in developing policy related to food issues. In *Food, People and Society: A European Perspective of Consumers' Food Choices* (L. Frewer, E. Risvik & H. Schifferstein, eds), pp. 415–432. Springer-Verlag, Berlin.

Rozin, P. (1989) The role of learning in the acquisition of food preferences by humans. In *Handbook of the Psychophysiology of Human Eating* (R. Shepherd, ed.), pp. 205–227, Wiley, Chichester.

Rozin, P. & Markwith, M. (1991) Cross domain variety seeking in human food choice, *Appetite* **16**, 57–59.

Scansaroli, J.A. & Szymanski, D.M. (2002) Who's minding the future? In *Retailing Issues Newsletter* **14**(1), Center for Retailing Studies, Texas A&M University.

Schlauser, E. (2000) *Fast Food Nation: What the All-American Meal is Doing to the World.* Penguin, London.

Shepherd, R. (1989) Factors influencing food preferences and choice. In *Handbook of the Psychophysiology of Human Eating* (R. Shepherd, ed.), pp. 25–56. Wiley, Chichester.

Shepherd, R. (2001) Does taste determine consumption? Understanding the psychology of food choice. In *Food, People and Society: A European Perspective of Consumers' Food*

Choices (L. Frewer, E. Risvik, & H. Schifferstein, eds), pp. 117–130, Springer-Verlag, Berlin.

Shiffman, L. & Kanuk, I. (2000) *Consumer Behaviour*. Prentice Hall, London.

Sobal, J. (2000) Sociability and meals: facilitation, commensality and interaction. In *Dimensions of the Meal: The Science, Culture, Business and Art of Eating* (H.L. Meiselman, ed.), pp. 119–133, Aspen, Gaithersburg, MD.

Solomon, M., Bamossey, G. & Askegaard P. (2002) *Consumer Behaviour: A European Perspective*, 2nd edn. Financial Times/Prentice Hall, Harlow.

Sullivan, O. (1997) Time waits for no (wo) man: an investigation of the gendered experience of domestic time, *Sociology* **31**(2), 221–240.

Tuck, M. (1976) *How Do We Choose?* Methuen, London.

Van Trijp, H.C.M. (1995) *Variety Seeking in Product Choice Behaviour*, Doctoral thesis, University of Wageningen.

Warde, A. & Hetherington, K. (1994) English households and routine food practises: a research note, *Sociological Review* **42**(4), 758–778.

Welsch, W. (1996) Aestheticization processes: phenomena, distinctions and prospects, *Theory Culture and Society* **13**(1), 1–24.

Worsley, T., Coonan, W. & Baghurst, P.A. (1983) Nice, good food and us: a study of children's food beliefs, *Journal of Food and Nutrition* **40**, 35–41.

Chapter 3

Perceived Risk and Product Safety in the Food Supply Chain

S. Kuznesof & M. Brennan

Objectives

(1) To contextualise the importance of public opinion in matters of food safety.
(2) To provide an overview of consumer safety in the food supply chain, focusing on consumer perceptions of food risks by giving theoretical and empirical insights into the ways in which individuals characterise and perceive food-related risks.
(3) To discuss the associated implications for the communication of food risks within the food supply chain.

Introduction

Public perceptions of risk and concerns about the safety of British food have resulted in significant changes in food safety management in the food supply chain and in the regulatory environment within which the food industry operates. By focusing on food safety and risk perceptions, this chapter complements the more generic concepts underpinning consumer decision-making and their applications to food choice as detailed in Chapter 2. Three main arguments developed in this chapter are: first, 'food safety' is a supply chain issue that needs to be understood by, and addressed by, all constituents including food producers, processors and manufacturers, wholesalers and retailers, and the catering industry. Second, consumers are arbiters of product success. They are heterogeneous in their needs, wants, attitudes and aspirations, and these require understanding and responding to appropriately. Thirdly, the communication of food risks to the public needs to be undertaken in an integrated manner, involving a dialogue between the food supply chain, the government and food consumers.

Thus, this chapter begins with an overview of the context in which food safety has become a food chain, government and public issue. Public perceptions of food safety are presented and contrasted with 'expert' beliefs. Theoretical and empirical research informing how individuals perceive food risks is described, and the implications of these analyses for the communication of food risks to the public are then discussed from psychological and behavioural perspectives. Two food crisis case studies, namely BSE (1986–99) and foot and mouth disease (FMD) (2001), are presented throughout the chapter to aid the discussions. Reference is also made to genetically modified foods (GM foods).

Background

Successions of food crises in the United Kingdom (UK) over the past two decades have placed food safety on the political, food supply chain, and public agenda. The consequences of food scares, in particular that of bovine spongiform encephalopathy (BSE) and its link to new variant Creutzfeldt–Jakob disease (vCJD) (Box 3.1), have resulted in declining public confidence and trust in the safety of food, the food industry, and the government's ability to adequately regulate, manage and communicate food risks. Political re-evaluation of food safety and the need to restore public faith in British food and its regulation (MAFF, 1998) resulted in the ministerial separation of public food safety interests from food production interests which were both held within the Ministry of Agriculture, Fisheries and Food (MAFF). One political outcome was the statutory creation in 2000 of a new consumer-focused institution, the Food Standards Agency (Food Standards Act, 1999), dedicated to taking a strategic view of the food chain, or 'farm to fork' approach, where food safety and consumer protection were founding aims (MAFF, 1998). The restructuring of the old MAFF into the newly created Department of Environment, Food and Rural Affairs followed shortly in June 2001.

In addition to these significant regulatory changes, primary producers in the food industry have undergone an enforced period of introspection of their primary production methods and practices following additional sectoral food crises such as swine fever in 1999 and FMD in 2001. Initiated by FMD, a review by the Commission on the Future of Farming and Food (2002) identified the need for the food industry, and in particular operators in the upstream agrifood supply chain, to become more market-oriented, by taking a 'whole food chain approach' to food production, processing, retailing and catering in a way that responds more proactively to customer needs and market trends, in particular by offering products with specific qualities and added value (Steenkamp, 1987; McInerney, 2000; Morris & Young, 2000; Weatherell *et al.*, 2002). This market-led behaviour has for some time been identified as an important factor in building a viable farming community (NFU, 1994; Ritson & Kuznesof, 1996), despite traditional government intervention in agricultural markets influencing farm business strategies. Such customer-focused visions are identified as crucial to commercial success (Peters & Waterman, 1982).

Indeed, the concept of 'marketing', whereby companies strive for efficiency and effectiveness by identifying, anticipating and satisfying customer needs thereby surpassing their competition (Barwell, 1965), is the accepted operating ethos in profit-seeking enterprises. Consumer power can be witnessed through the thousands of individual purchasing decisions that can determine the success or failure of a

Box 3.1 Characteristics of the BSE/new variant CJD food crisis.

Relevance/notoriety	The bovine spongiform encephalopathy (BSE)/new variant Creutzfeldt–Jakob disease (vCJD) crisis was notable for its longevity, spanning over a decade from 1986, and for the failure in government and industry management of risks when faced with uncertainties in the risk assessment process.
Impact of crisis	In 1996 alone, beef consumption fell by 11% throughout the European Union and cost US$5 billion in subsidies (Anon., 1997). The severely diminished public confidence in the government initiated UK and EU political institutional restructuring of public health protection. This announcement has been described as 'the culmination of . . . years of mismanagement, political bravado and a gross underestimation of the public's capacity to deal with risk' (Powell & Leiss, 1997).
Notification of crisis	Cases of BSE in cattle were first identified by the Veterinary Laboratories Agency, Weybridge, UK, in November 1986, and in 1988 it was made a notifiable disease (Ministry of Agriculture, Fisheries and Food, 2000; Food Standards Agency, 2000). Despite public concern about the potential link of BSE with CJD, it was not until March 1996 that Health Secretary Stephen Dorrell (Dorrell, 1996) made a statement to the House of Commons presenting evidence of a new variant CJD and not ruling out a link with the consumption of beef from cattle with BSE (Food Standards Agency, 2000).
Description of hazard	BSE is a slowly progressing, fatal nervous disorder of adult cattle that causes a characteristic staggering gait (hence usage of the term 'mad cow disease' to describe it) (Ministry of Agriculture, Fisheries and Food, 2000). The human equivalent of the disease, Creutzfeldt–Jakob disease, a neurological disease affecting the brain, occurs in one person in a million around the world. With an incubation period of up to 30 years, the disease has historically been associated with the elderly. Once the symptoms present themselves, the infected individual rapidly degenerates and usually dies within six months of diagnosis. There is at present no known cure for the disease.
Description of risk	The cause of BSE was hypothesised to stem from ruminant derived meat and bone meal, and therefore in 1988 a ban was placed on the sale or supply of feeding stuffs for ruminants that included ruminant protein (Food Standards Agency, 2000). There was little known about the precise nature of the infective agent, or the routes of transmission within and between species, the extent to which the infective agent could cross the species barrier, the spread of the agent within the human body, and the variation in individual susceptibility. Following a number of cases of a previously unrecognised disease pattern consistent with CJD, the National Creutzfeldt Jakob Disease Surveillance Unit (CJDSU) concluded that, in the absence of any credible alternative, the most likely cause of new variant CJD (vCJD) disease was exposure to a BSE-like agent before the introduction of a specified bovine offal ban in 1989.

product item. For example, in the USA and UK, 75% to 85% of new products fail to maintain a retail presence beyond one year (Buisson, 1995). Public rejection can limit the scope or adoption of new technological innovations; for example, food irradiation (Henson, 1995) and GM foods (OECD, 1989). Indeed, changes to the purchasing environment such as the threat of a 'food scare' can temporarily alter food-purchasing patterns. In the case of the *Salmonella* in eggs scare in 1988, public perceptions of risk associated with contracting salmonellosis from egg consumption cost the industry £70 million in lost income.

Thus, in order to anticipate and respond to consumers' needs and wants effectively, the food supply chain needs to understand food consumers' perceptions and attitudes to food safety.

Consumer perceptions of food safety

A number of surveys and opinion polls have sought to identify consumer attitudes to food and its safety (see, for example, Frewer *et al.*, 2003). Table 3.1 typifies the key concerns identified in exploratory focus group research (Frewer *et al.*, 2001). An asterisk denotes the discussion of a food concern in each focus group. These concerns represent issues most salient to individuals within the groups. The research consisted of nine group discussions including two experimental groups that were held over a three-month period from September to December 1999. Group profile selection reflected an intention to interview contrasting gender, social class and age profiles. The female gender bias (33 of the 46 participants in the seven main groups) was to reflect the dominance of women as the 'gate-keepers' of food selection (as they remain principal purchasers of food). The food concerns have been categorised according to hazard type. The results show a range of concerns varying from anxieties relating to each stage of the food chain; for example, the contents of animal feeds at the input supply end of the food chain, to the management of food safety. Specific food safety issues, such as GM foods and the use of additives and preservatives in processed foods, were also frequently mentioned. Table 3.1 also highlights the heterogeneity in attitudes to food concerns. Not only were there differences in the degree of importance attached to various food concerns, there were also differences in the prioritisation of food as an issue of concern. For example, amongst low-income consumers, food safety was a low priority issue when placed in the context of wider life concerns such as housing and health. This lack of homogeneity amongst the public can be seen in, for example, the case of attitudes to GM foods. Historically, segmentation studies that categorise consumers into groups according to certain criteria have used conventional demographic and socio-economic criteria, such as gender, income and socio-economic status. Exploratory research identified four typologies of consumer (Kuznesof & Ritson, 1996). The groupings were made according to similarities in the 'values' of the respondents (Box 3.2).

Notwithstanding individuals' concerns about food and its safety, absolute food safety, although a preference, is not an expectation, as personal experience of food

Table 3.1 Unprompted public concerns about food and its safety (Frewer *et al.*, 2001).

Food concerns	**Group**						
	1	**2**	**3**	**4**	**5**	**6**	**7**
Animal Health and Welfare							
Salmonella in eggs			*		*	*	*
BSE/vCJD	*	*	*	*	*	*	*
Animal welfare	*		*		*		
Animal products in animal foods		*	*	*	*	*	*
Hygiene							
Hygiene of food preparation premises			*	*			
Hygiene of domestic food storage and preparation					*		
Contaminants and Residues							
Pesticides	*	*				*	*
Antibiotics			*		*		*
Hormones	*	*					*
Radiation							
Chernobyl (lamb and milk)			*				
Novel Foods							
GM foods	*	*	*	*		*	*
Additives, Flavourings							
Additives	*		*		*	*	*
E numbers	*		*		*	*	
Irradiation							
Irradiation	*		*				
Decision-Making Process (regulatory)							
Information suppression	*	*			*	*	*
Lack of information (available to government)			*				
Nutrition and Health							
Fast food	*				*	*	*
Children having a balanced diet	*	*					
Nutrition in general	*	*					*
Dieting/weight management	*						
Nutritional content of food		*				*	
Food Business							
Profit motive usurping consumer safety	*	*		*	*	*	*
Information Provision							
Contradictory information	*	*		*		*	*
Media reporting of food scares	*			*	*	*	*
Food Attributes							
Poor flavour (fruit)			*				
Poor labelling (origin)			*				
Food decaying before sell-by-date					*		
Personal							
Lack of nutrition education	*	*					*
Feeding fussy children							*
Affording good food		*		*	*		*
Taking the right course of action (uncertainty)			*	*			
Lack of information	*		*				*

Box 3.2 GM food consumer typologies (Kuznesof & Ritson, 1996).

Typology	Description
Refuser	A small group who rejected the use of gene technology in food production and, by implication, GM foods. The reasons for this were based on an instinctive or intrinsic hostility to gene technology, moral or religious objections such as 'tampering with nature' or 'playing God' and also concerns about the safety and control of the technology.
Traditional trier	Typified by economically vulnerable consumers with low disposable incomes for whom 'price' was central to food purchasing decision-making. If GM foods were cheaper than non-GM alternatives, the GM offering would most probably be purchased.
Enthusiastic trier	Displayed 'innovator' or 'pioneer' characteristics and were more predisposed to the benefits of the technology, believing it to have a role in economic and personal progress. For this group concerns of safety were relegated to regulating authorities and ultimate acceptance was expressed as the degree of satisfaction arising from the trial of a GM food.
Undecided	Constituted the majority of the discussants. For the many members of this group, the decision to accept or reject GM foods was dependent upon many factors such as beneficiary, nature of modification, type of product modified (e.g. plant, fish or mammal), etc.

crises attests (Frewer *et al.*, 2001). This has implications for the communication of food risk information throughout the food supply chain and amongst risk regulators.

Comparison of lay and expert perceptions of food risks

'Experts' and the public often rank the relative importance of risks differently (Sandman, 1987; Slovic, 1987). Box 3.3 is a compilation of such rankings and highlights an almost inverse relationship between expert and scientific perceptions of food risks.

Box 3.3 Expert and public risks associated with food ranked in order of importance (Smith, 1997).

Expert/Scientific	Public
(1) Microbial contamination	(1) Food additives
(2) Nutritional imbalance	(2) Pesticide residues
(3) Environmental concerns	(3) Environmental concerns
(4) Natural toxicants	(4) Nutritional imbalance
(5) Pesticide residues	(5) Microbial contamination
(6) Food additives	(6) Natural toxicants

Table 3.2 UK deaths per year related to diet and food (Krebs, 2001).

Risk	Number of deaths
Cardiovascular disease (CVD)[a]	73 000
Cancer[a]	34 000
Food-borne illnesses (estimated)	50
Food allergy	20
vCJD	15
Genetically modified organisms, pesticides, growth hormones	0

[a] Assumes one-third and one-quarter of total CVD and cancers, respectively, are diet related.

If the actual number of UK deaths per year related to diet and food is examined (Table 3.2), it can be shown that the expert perceptions of risk reflect more accurately the 'real' food risks rather than public 'perceptions' of food risks. Those risks ranked as most important by the public have resulted in very few, if any, deaths among the public. The risks ranked as most important by experts contribute to a significant number of food-related deaths per year in the UK; for example, death from cancer and heart disease has been shown to be strongly linked with a nutritionally imbalanced diet.

This position has led some experts to believe that the public is incapable of dealing with the concept of food risk (Fessenden-Raden *et al.*, 1987); this attitude fostered the development of the so-called 'deficit model' of communicating risks to the public. The 'deficit model' has its origin in the public understanding of science movement, where it was assumed that the public was in some way ignorant of the scientific 'truth' about technical risk assessment and probabilities. The model sought to 'rectify' the knowledge gap between the originators of scientific information (the scientific elite) and the subsequent literacy of the audience or recipients of the information (Higartner, 1990). Rather than reducing the knowledge gap, however, this approach served to reinforce the differences between scientific experts and the general public. Despite attempts to move away from the deficit model approach to risk communication, recent research for the Food Standards Agency examining the communication of the concept of 'uncertainty' in food risk communication has highlighted that many members of the scientific community still ascribe to it (Frewer *et al.*, 2001). Amongst the sample of scientific experts, there was widespread agreement that the public would be unable to cope with risk uncertainty information and that providing information about scientific uncertainty would have a negative impact on the extent to which the public trust science, scientists and scientific institutions (Frewer *et al.*, 2002).

By comparison, consumer attitudes towards food risk uncertainty found the reverse to be true, i.e. the public were accepting of the concept of uncertainty, wanted to be informed of uncertainty in food risk information, and admissions of uncertainty by the government were likely to increase the credibility of the message and trust in the information source (Frewer *et al.*, 2001). Some of the differences in

Box 3.4 Some characteristics of the two languages of risk communication (Powell & Leiss, 1997).

'Expert' assessment of risk		**'Public' assessment of risk**
Scientific		Intuitive
Probabilistic	*Barriers*	Yes/no
Acceptable risk	*to*	Safety
Changing knowledge	*mutual*	Is it or isn't it
Comparative risk	*understanding*	Discrete events
Population averages		Personal consequences
A death is a death		It matters how you die

expert and public assessment of risk that have fostered barriers to mutual understanding are highlighted in Box 3.4. For example, experts define risks in the terminology and procedures of science itself, with a detached objectivity relating to probabilistic risk assessments and total numbers of individuals likely to be affected (Groth, 1991). By contrast the lay public are more intuitive in the assessment of risk, drawing from personal experience, using a more qualitative language to describe the risks and making attempts to relate the risks to their individual situations.

Psychological theories of perceived risk

To explain 'why' individuals have particular beliefs about food risks, two theoretical approaches are presented that give insights to the subject. Risk perceptions are socially constructed and individual responses to risk are driven by the individual's beliefs and perceptions, and not (as highlighted above) by the scientifically based technical risk estimates of experts (Frewer, 1999).

Within social psychology, substantial research has been undertaken to help understand the factors affecting the public's perceptions of risk and the context in which they are created. Research that seeks to explain why some risks invoke more alarm, outrage, anxiety or dread than others regardless of scientific estimates of their seriousness is referred to as the 'psychometric paradigm'.

The psychometric paradigm, which originated in the work of Starr (1969) and was explored principally by Slovic and colleagues (Slovic, 1987, 1992; Fischoff, 1995), was developed in order to understand the psychological characteristics which underlie an individual's response to risk. It is founded on the assumption that risk itself is a subjectively defined quality, and socially constructed in that it is influenced by a variety of factors: psychological, institutional, social and cultural (Frewer *et al.*, 2001). It is these factors that may be important determinants in lay responses to risk information and thus explain the differences in scientific and lay perceptions of risk (and indeed cultural differences in risk perceptions). Covello (1992) has identified 47 factors that influence the public's perception of risk. Box 3.5 provides a summary of these key 'risk amplification' or 'dread' or 'fright' factors (factors that are believed to pose a greater threat than technical risk estimates would suggest) and 'risk attenuation' or 'comfort' factors (factors that reduce perceptions of

Box 3.5 Risk amplification and risk attenuation factors (Bennett, 1999).

Risk amplification factors	Risk attenuation factors
Risk is involuntary	Risk is voluntary
Third party control	Individual control
Inequitable	Equitable
Inescapable	Avoidable
Unfamiliar or novel	Familiar
Man-made	Natural
Effects unknown	Effects known
Long term effects	Short term effects
Irreversible damage	Damage is reversible
Danger to vulnerable groups or future generations	Population equally affected
Risk poorly understood by scientists	Well understood by scientists
Contradictory statements from responsible sources	Consistent statements from responsible sources

risk) that are influential in determining public responses to risk. These fright factors can help practitioners understand the ingredients of public concerns about risk and to calibrate which risks might be perceived as arresting or alarming (Bennett, 1999).

Certain hazards are likely to be amplified by the general public whilst others are attenuated (Kasperon *et al.*, 1988; Flynn & Slovic, 1995). The degree of public concern created by a risk will, in general, be amplified in situations where expert institutions are suspected of either withholding information about food risks or side stepping the question of accountability. To illustrate this, the case of GM foods can be analysed. Public disquiet regarding gene technology and genetically modified foods could possibly have been anticipated if the nature and characteristics of gene technology were considered in light of the risk amplification factors given in Box 3.5.

Gene technology is a science-driven (rather than market-led) technology, a 'man-made' means of production that can enable the transfer of genetic material across species boundaries, a phenomenon that would not occur in nature. This complex and novel technology is suggested to have some long term and irreversible effects including, for example, on bio-diversity through the breeding of GM plant strains with conventional counterparts through 'genetic drift' (Hatchwell, 1993). As a technology in its infancy, the potential risks have been publicly debated, with contradictory evidence presented by opponents and proponents. In addition, without satisfactory labelling of GM foods, segregation of ingredients, or identification of GM ingredients, individuals who wish not to purchase or eat GM foods may involuntarily be doing so, unable to exercise personal control in their purchasing decisions.

Following the public's experience of BSE in the UK, which provided a 'conceptual template' or experience of regulatory capabilities in the face of uncertainty and incomplete information, the public's general risk perceptions have been sufficiently significant to motivate major grocery chains to limit or halt the sale of GM foods. The social amplification of risk, as described through this case of GM foods,

Box 3.6 Characteristics of the foot and mouth disease food crisis.

Relevance/notoriety	The foot and mouth disease (FMD) crisis critically highlighted the role, behaviour and practices of the UK food supply chain in terms of, for example, the excessive transportation of animals around the country, the pressure from supermarkets for the centralisation of animal slaughter and the illegal importation of meat into the UK which was believed to be the cause of the crisis (Harvey, 2001). The disease highlighted not only the integrated relationship between each member of the food supply chain but also the inter-relationship of farming in the wider rural economy.
Impact of crisis	Estimated that approximately 5.5 million animals in total were slaughtered. The financial implications of the crisis both from the farming and wider rural community have been estimated to have cost the overall UK economy in the region of £7 billion, with approximately £750 million being spent on the control and eradication of the disease (Harvey, 2001).
Notification of crisis	The outbreak of FMD in the UK was confirmed on 20 February 2001, after cattle and pigs at an abattoir in Essex were diagnosed with the disease (Lowe *et al.*, 2001; DEFRA, 2002). The source of the disease was traced to a pig-fattening unit 400 km away at Heddon on the Wall, Northumberland. Unfortunately, by the time the source had been traced, the disease had already spread rapidly to farms in the south-west, north-west and Midlands of England, Wales and southern Scotland (Lowe *et al.*, 2001).
Description of hazard	FMD posed no health hazard to the public: the disease affected only animals. It affects the condition of adult animals, including milk yields in dairy cows. It can be fatal for unborn and young animals.
Description of risk	FMD is both seriously debilitating and exceptionally contagious, spreading both by direct and indirect contact (Harvey, 2001). It affects cloven-hoofed animals, in particular cattle, sheep, goats and pigs. FMD cannot be passed on to humans through the consumption of products from infected animals (Harvey, 2001).

demonstrates the importance of effective risk communication, especially in cases where responses to hazards are likely to be out of proportion to the risks they pose.

Conversely, the experience of the FMD crisis (Box 3.6) showed that 'consistent statements from responsible sources' provided reassurance to the public, following early announcements of the presence of FMD amongst UK ruminant herds, and helped allay public fears that the disease may be transmissible to humans. The unified public announcements by the Chief Veterinary Officer, the Chief Medical Officer and the Food Standards Agency clearly stating the disease was not a threat to human health meant that while FMD remained a high profile farming industry and rural economy crisis, it did not at any stage become a public health crisis.

Behavioural theory of perceived risk

The preceding discussion analysed psychological research on risk perceptions. In this section, individual purchasing decisions or the consumer behaviour approach to the 'theory of perceived risk' are presented.

Consumer behaviour is commonly understood as the study of 'the activities involved in obtaining, consuming and disposing of products and services, including the decision processes that precede and follow these actions' (Engel *et al.*, 1995). In an attempt to understand the factors affecting these activities, models of food choice typically combine factors relating to the individual, the product under consideration and the purchasing environment. All forms of consumer behaviour have been described as 'risk-taking' behaviour to the extent that the consequences of any purchasing or consumption action cannot be foreseen with complete certainty (Bauer, 1967). Seven types of 'risks' or 'losses' have been associated with the processes of purchasing, consuming and disposing of products. These 'risks' or losses are physical, performance, social, psychological, financial, time and opportunity. Research indicates that individuals' perceptions of risk and their subsequent purchasing behaviour are causally linked, with risk perceptions an important explanatory variable of the latter (Mitchell & Greatorex, 1990; Huang, 1993; Eom, 1994). However, as with all forms of consumer behaviour, risk is temporal, being associated with a particular product, unique to a particular person at a particular point in time. For example, in 1988, egg purchases were likely to be a part of routinised purchasing behaviour for many consumers until information about the presence of *Salmonella* in eggs prompted a re-evaluation of their egg purchasing behaviour. From a product perspective, there are different types of buying decision based on the level of consumer involvement and the differences across brands or product lines. For some individuals, food is an important life consideration and significant time may be dedicated to selecting and purchasing foods appropriate to beliefs or lifestyle. For example, an individual keen to manage weight may spend time seeking low-fat food alternatives to conventional counterparts. For others, food is not as significant, with other factors such as health and housing more important to them. For these, often lower-income, consumers, price is a major determinant in a food purchasing decision.

There are many factors affecting individual differences in approaches to food safety. Some individuals exhibit both risk-averse and risk-seeking behaviour across a wide variety of situations (Kahneman & Tversky, 1979). Consumers have inherent predispositions to seek or avoid risk in purchase situations (Dowling, 1986). Depending upon an individual's tolerance to risk, there may be situations under conditions of boredom, curiosity or variety seeking, where a less 'risky' product may be rejected in favour of a more 'risky' product. To the extent that consumers may meaningfully order products with respect to perceived riskiness, trade-offs between product purchases can be made according to the benefits sought.

Consumers' perceptions of risk therefore stimulate information search and risk handling. When faced with a potentially risky purchasing decision, consumers may attempt to reduce the risk involved by developing strategies to reduce perceptions of risk and enable them to act with relative confidence in uncertain situations. Four generic strategies to resolve or reduce perceived risk are the following (Roselius, 1971):

(1) Reduce the perceived uncertainty about the product, or reduce the severity of real or imagined loss suffered if the product does fail.
(2) Shift from one type of perceived loss towards one for which there is more tolerance.
(3) Postpone the purchase.
(4) Make the purchase and absorb the unresolved risk.

Particular behaviours drawn from consumers' experiences of purchasing beef during the BSE/vCJD food crisis are highlighted in Box 3.7 together with a number of 'risk relieving' measures to eliminate the risk, or reduce or eradicate the presence of a hazard, or maintain the existing food purchasing patterns. These behaviours were all based upon some appraisal of food risk information, experience of previous food scares, trust in beef retailers, trust in information provided by government, and other individual differences such as age and life cycle stage (older people were less concerned about changing beef purchasing patterns than parents of young children).

Box 3.7 Food selection reactions to BSE (Frewer *et al.*, 2001).

Behaviour change	Rationale
Stop purchasing beef	Eliminate the risk of contracting CJD within the family unit, by withdrawing from the diet entirely the food to which the hazard was attached.
Modify beef purchases	(1) Reduce the quantities purchased of the food to which the hazard is attached, thereby reducing the risk of contracting CJD. (2) Improve the quality of the food purchased in an attempt to reduce or eradicate the presence of the hazard (as it is understood), and therefore the risk by, e.g. — trading-up in quality by buying more expensive joints of beef rather than burgers — purchasing organic beef — purchasing Scottish instead of English beef — purchasing breed specific beef such as Aberdeen Angus.
No behaviour change	(1) Individuals do not believe they are at risk or the risk is infinitesimal and — they purchase beef from trusted suppliers — they enjoy eating beef, or (2) Individuals believe they have been exposed to the hazard and any behaviour change will be meaningless.
Purchase more beef	The lower price of beef twinned with attitudes described under 'no behaviour change' made the purchase of beef attractive.

Risk relieving strategies can also be initiated by sellers, and consideration of incorporating risk reduction or risk management strategies within the food companies' marketing mixes is recommended (Yeung & Morris, 2001). Examples of such strategies to reduce perceptions of food safety include quality assurance schemes, labelling and product information (for further examples, see Yeung & Morris, 2001).

Discussion

It has been suggested in this chapter that public concerns about food safety are a supply chain issue. Consumers' perceived risks associated with food cover the length of the food supply chain, and public anxieties about food safety have had severe economic consequences on industries at the centre of food scares. The general consensus from both within and outwith the food supply chain is to be more market-oriented in approach to the final consumer, looking beyond the needs of immediate customers in the food chain and incorporating the needs of end users or food consumers. However, the approach to the final consumer needs to be integrated. For example, although individual companies can incorporate risk reducing strategies into their marketing mixes, replications of isolated initiatives can be confusing to the public. For example, the Food Standards Agency has recommended a re-evaluation of quality assurance schemes which have proliferated in the wake of food scares and are potentially confusing to the public.

The heterogeneity of the UK population has been clearly demonstrated in this chapter. This heterogeneity poses an immense challenge to the whole food supply chain. Recognition by the food supply chain that the UK public have a variety of beliefs, attitudes and concerns about food, in particular with relation to how food is produced and its safety, is vital to developing a more integrated, market oriented, customer focused food supply chain. It is no longer an option to base communication, education, product development and risk reducing strategies on the inherent beliefs and values held by the industry, which in most cases have been shown not to correspond to those of the general public. The food supply chain needs to embrace the general public and make them a vital link in the food chain if a more market oriented, customer focused and potentially more effective and efficient food supply chain is to be created in the UK.

Further research conducted for the Food Standards Agency (Frewer *et al.*, 2003) has indicated that not only do the UK public hold a range of different concerns about food and its safety, but also that the range of concerns and the level of concern expressed varies between different groups within the population. For example, young educated males are not very concerned about anything to do with food and its safety. These results have huge implications for the way the food supply chain delivers and communicates information about food and its safety to the UK public. The method and format of delivery may need to be tailored to meet the differing needs of the various groups within the population, to ensure that all people receive the necessary information, in particular when there are food scares that have health implications for either particular groups or the population as a whole.

In conclusion, it is imperative that members of the UK food supply chain, from producers to retailers, educate themselves about the public and how they behave under conditions of risk and uncertainty. The public will not react or behave as a homogenous whole; differing response behaviours and coping strategies will be exhibited, all of which must be recognised as valid. Dismissing these differing behaviours as irrationality, as was the case during the BSE crisis, is no longer acceptable. It is imperative that members of the food supply chain work more closely with the regulatory bodies, including the Food Standards Agency, to ensure that the public are fully informed about all issues concerning the food they eat, especially during times of crisis where there is a potential risk to public health. In addition, it is vital that precautions are taken to reduce the likelihood of amplifying the risk through poor uninformed communication strategies. The lack of adequate and understandable information in the past, especially during the BSE crisis, has led to the food supply chain, the government and the various regulatory agencies losing the trust and credibility of the UK public. Improved information flow both between the members of the food supply chain and the public will help begin to rebuild the damage that has been caused by years of food scares and inadequate communication and information provision. By adopting a fully integrated, transparent and honest approach to communicating information about food and its safety to the public, the food supply chain will begin to transform itself from what is currently quite a disjointed and inharmonious set of individual stakeholders into an effective and efficient body.

Acknowledgements

Selected research findings reported within this chapter stem from data emanating from Food Standards Agency funded project no. DO1005 'Communicating Risk Uncertainty with the Public' (Frewer *et al.*, 2001). The authors gratefully acknowledge the contributions of co-researchers in this project and also the assistance of Chris Ritson for reviewing drafts of this chapter.

Study questions

1. How do the characteristics of food risk and risk regulators affect consumers' perceptions of food safety?
2. How can knowledge of psychological and behavioural theories of risk assist in the management and communication of food crises?
3. What guidance would you give to members of the food supply chain in communicating food risks to the public?

References

Anon. (1997) How BSE crisis forced Europe out of its complacency. Nature **2** January.

Barwell, C. (1965) The marketing concept. In *The Marketing of Industrial Products* (A. Wilson, ed.). Hutchinson, London.

Bauer, R.A. (1967) Consumer behaviour as risk taking. In *Risk Taking and Information Handling in Consumer Behaviour* (D.F. Cox, ed.), Harvard University Press, Boston.

Bennett, P. (1999) Understanding responses to risk: some basic findings. In *Risk Communication and Public Health* (P. Bennett & K. Calman, eds), Oxford University Press, Oxford.

Buisson, D. (1995) Developing new products for the consumer. In *Food Choice and the Consumer* (D.W. Marshall, ed.), pp. 182–215. Blackie Academic and Professional, London.

Commission on the Future of Farming and Food (2002) *Report of the Commission on the Future of Farming and Food: Farming and Food. A Sustainable Future*. The Stationery Office, London (http//www.cabinet-office.gov.uk/farming).

Covello, V. (1992) Risk communication: an emerging area of health communication research. In *Communication Yearbook* (S. Deetz, ed.), pp. 359–373. Sage, Newbury Park and London.

DEFRA (2002) *Origin of the UK Foot and Mouth Epidemic in 2001*. The Stationery Office, London (http//www.cabinet-office.gov.uk).

Dorrell, S. (1996) Statement in Parliament, 20 March 1996.

Dowling, G.R. (1986) Perceived risk: the concept and its measurement. *Psychology and Marketing* **3**, 193–210.

Engel, J.F., Blackwell, R.D. & Miniard, P.W. (eds) (1995) *Consumer Behaviour*, 5th edn, Dryden Press, London.

Eom, Y.S. (1994) Pesticide residue risk and food safety valuation: a random utility approach. *American Journal of Agricultural Economics* **76**(4), 760–772.

Fessenden-Raden, F., Fitchen, J.M. & Heath, J.S. (1987) Providing risk information in communities: factors influencing what is heard and accepted. *Science, Technology and Human Values* **12**, 94–101.

Fischoff, B. (1995) Risk perception and communication unplugged – 20 years of process. *Risk Analysis* **15**(2), 137–145.

Flynn, J. & Slovic, P. (1995) Nuclear wastes and public trust. *Forum for Applied Research and Public Policy* **14**, 92–101.

Food Standards Act (1999) The Stationery Office, London.

Food Standards Agency (2000) *Review of BSE Controls*. The Stationery Office, London.

Frewer, L.J. (1999) Public risk perceptions and risk communication. In *Risk Communication and Public Health* (P. Bennett & K. Calman, eds), Oxford University Press, Oxford.

Frewer, L.J., Hunt, S., Miles, S., Brennan, M., Kuznesof, S., Ness, M. & Ritson, C. (2001) *Communicating Risk Uncertainty with the Public*. Final Report & Technical Annex for the UK Food Standards Agency. Project DO1005, Food Standards Agency, London.

Frewer, L.J., Miles, S., Brennan, M., Kuznesof, S., Ness, M. & Ritson, C. (2002) Public preferences for informed choice under conditions of risk uncertainty. *Public Understanding of Science* **11**, 363–372.

Frewer, L.J., Hunt, S., Kuznesof, S., Brennan, M., Ness, M. & Ritson, C. (2003) The views of scientific experts on how the public conceptualise uncertainty. *Journal of Risk Research* **6**(1), 75–85.

Groth, E. (1991) Communicating with consumers about food safety and risk issues, *Food Technology* **45**(5), 248–253.

Harvey, D.R. (2001) *What Lessons from Foot and Mouth? A Preliminary Assessment of the 2001 Epidemic*. Working Paper 63, University of Newcastle upon Tyne, Centre for Rural Economy, Newcastle upon Tyne.

Hatchwell, P. (1993) Opening Pandora's box: the risks of releasing genetically engineered organisms. *Ecologist* **19**(4), 130–137.

Henson, S. (1995) Demand-side constraints on the introduction of new food technologies: the case of food irradiation. *Food Policy* **2**(2), 111–127.

Higartner, S. (1990) The dominant view of popularisation: conceptual problems, political uses. *Social Studies of Science* **20**, 519–539.

Huang, C.L. (1993) Simultaneous equation model for estimating consumer risk perceptions, attitudes and willingness-to-pay for residue-free produce. *Journal of Consumer Affairs* **27**(2), 377–388.

Kahneman, D. & Tversky, A. (1979) Prospect theory: an analysis of decisions under risk, *Econometrica* **47**, 263–291.

Kasperon, R.E., Renn, O., Slovic, P., Brown, H.S., Ernel, J., Goble, R., Kasperson, J.X. & Ratick, S. (1988) The social amplification of risk: a conceptual framework. *Risk Analysis* **8**, 177–187.

Krebs, J. (2001) *Is Food Safe*? Public Lecture, University of Newcastle upon Tyne.

Kuznesof, S. & Ritson, C. (1996) Consumer acceptability of genetically modified foods with special reference to farmed salmon. *British Food Journal* **98**(4/5), 39–47.

Lowe, P., Edwards, S. & Ward, N. (2001) *The Foot and Mouth Crisis: Issues for Public Policy and Research*. Working Paper 64, Centre for Rural Economy, University of Newcastle upon Tyne, Newcastle upon Tyne.

McInerney, J. (2000) Re-orientating UK agriculture. *Farm Management* **11**(4), 217–231.

Ministry of Agriculture, Fisheries and Food (1998) *The Food Standards Agency: A Force for Change*. Document Cm 3830. The Stationery Office, London.

Ministry of Agriculture, Fisheries and Food (2000) *The BSE Report*: *Findings and Conclusions*, Vol. 1. The Stationery Office, London.

Mitchell, V.W. & Greatorex, M. (1990) Consumer perceived risk in the UK food market. *British Food Journal* **92**(2), 16–22.

Morris, C., & Young, C. (2000) 'Seed to shelf', 'teat to table', 'barley to beer' and 'womb to tomb': discourses of food quality and quality assurance schemes in the UK. *Journal of Rural Studies* **16**, 103–115.

NFU (1994) *Food from the Countryside*. National Farmers' Union, London.

OECD (1989) *Biotechnology, Economic and Wider Impacts*. Organisation for Economic Co-operation and Development, Paris.

Peters, T.J. & Waterman, R.H. (1982) *In Search of Excellence: Lessons from America's Best Run Companies*, Harper and Row, New York.

Powell, D.A. & Leiss, W. (1997) *Mad Cows and Mother's Milk. The Perils of Poor Communication*. McGill, Queen's University Press, London.

Ritson, C. & Kuznesof, S. (1996) The role of marketing rural food product. In *The Rural Economy and the British Countryside* (P. Allanson & M. Whitby, eds), Earthscan Publications, London.

Roselius, T. (1971) Consumer ranking of risk reduction methods. *Journal of Marketing* **35**(1), 56–61.

Sandman, P.M. (1987) Risk communication: facing public outrage. *EPA Journal* **13**(9), 21–22.

Slovic, P. (1987) Perception of risk. *Science* **236**, 280–285.

Slovic, P. (1992) Perceived risk, trust, democracy. *Risk Analysis* **13**(6), 675–682.

Smith, J. (1997) *The New European Food Safety Policy to Promote Good Health.* Club de Bruxelles, Brussels.

Starr, C. (1969) Social benefit versus technological risk. *Science* **165**, 1232–1238.

Steenkamp, J.-B. (1987) Conjoint measurement in ham quality evaluation. *Journal of Agricultural Economics* **38**, 473–480.

Weatherell, C., Tregear, A. & Allinson, J. (2002) *How Responsive are Consumers to 'Buying Local'? A Study of Public Perceptions of Food and Farming in the Wake of Foot and Mouth Disease.* Paper presented at Agricultural Economics Society Annual Conference, University of Wales, Aberystwyth, 8–11 April 2002.

Yeung, R.M.W. & Morris, J. (2001) Food safety risk: consumer perception and purchase behaviour. *British Food Journal* **103**(3), 170–187.

Chapter 4

Procurement in the Food and Drink Industry in the early 21st Century

J. Allinson

Objective

(1) To develop the reader's awareness and understanding of the strategic importance of procurement in the food and drink supply chain.

Introduction: definition of procurement

The words purchasing and procurement are often used interchangeably to define the action of buying or obtaining goods and/or services by effort from a third party to make them available within an enterprise or organisation to allow fulfillment of business objectives in a timely and cost efficient manner (Steele & Court, 1996; Steward, 1997). Purchasing and procurement are in fact very different: though both words describe the act of acquisition, procurement is a more strategic way of acquiring goods and services to help enterprise-wide strategies and objectives to be met.

This chapter therefore relates to procurement. The internal and external environment in which a business operates affects procurement decision-making. Steward (1997) has highlighted that the environment is complex and that decision making within it is affected by the following:

- the size of the organisation that is buying or selling;
- the relative complexity of a required input relative to end product and geographical area covered;
- public and private sector regulations and rules;
- the type of company involved (whether it is a service or a goods provider);
- the type of companies involved and end market (whether they are wholesale, retail or service oriented); and,
- the variety and type of raw materials supplier.

Procurement is clearly therefore a core business function at the interface between a business organisation and the external environment in which it trades and operates. The procurement function therefore involves complex and crucial decision-making. The chapter draws from a variety of relevant literature to illustrate how the approach to and style of purchasing and procurement management has evolved between the mid 20th century and the early 21st century and to describe how this has affected the UK food and drink industry. To do this, the next section briefly describes how the role, function and treatment of procurement has evolved, whilst the different management approaches that aim to provide efficient and consumer demand responsive procurement and supply chains that have prevailed since the 1980s are examined in the third section. This is followed by a summary of the key concepts and issues associated with these approaches.

Role, function and approaches to purchasing and procurement in western business between the mid-20th and early 21st centuries

In the 1960s, purchasing and procurement was very much a clerical or administrative job; in the 1980s it became a way to achieve commercial advantage; in the 1990s through to the early 21st century it has served a strategic role and function in business organisations (Nellore & Soderquist, 2000; Steele & Court, 1996; Steward, 1997; Quayle, 2002). This section describes why this was the case and the way in which purchasing and procurement was conducted as a result.

Fordism and post-Fordism

Between the mid-20th century and the late 1970s, production and supply were oriented around the development of commodity-like products for mass market demand, using what has been referred to as a 'Fordist approach' (Warde, 1997). During this period, the procurement function was an administrative task based upon the use of adversarial and competitive approaches in horizontal and vertical supply linkages (Saunders, 1997; Virolainen, 1998). In the food and drink industry, such objectives were translated into a need to secure as much of a resource as possible at the lowest possible cost for mass processing into largely undifferentiated finished product ranges that were marketed in small grocery shops and supermarkets.

By the end of the 1970s, during a period of global economic recession and upheaval, market supplies began to far exceed demand, and the need to be market and consumer demand responsive began to influence production and supply to encourage growth of the economy. Procurement thus gained strategic relevance because of its role as the interface between a business and its suppliers. The role of procurement consequently changed from an administrative task or mechanism into a core business function charged with ensuring that inputs to an enterprise or organisation were available at the right time and cost, and, most importantly, were

Box 4.1 Key characteristics of food consumption in the UK from the late 1970s to the early 21st century (adapted from Tansey, 1994; Warde, 1997; Atkins & Bowler, 2001).

(1) Increased demand for chilled and fresh food products with a short shelf life.
(2) Shorter food product life cycles and the growth of 'own-label' food products designed to meet the rapid changes in consumer demand.
(3) Incentives for all parties to squeeze costs in the supply chain to gain additional margin and enhanced competitiveness.
(4) The internationalisation of the food supply chain to meet perennial demand for particular foods.
(5) Persistent and substantial demand for food at the 'right price': the continued existence of the price conscious consumer.

able to meet market and consumer demand. This approach to business operations has been referred to as post-Fordist.

The need for market responsiveness has, in the food industry, revealed a wide range and variety of food consumer demands. This subject area has been, and continues to be, investigated and explored in consumer behaviour, marketing, economic, and geographical academic work. Box 4.1 above draws from such work and summarises how this has impacted the food and drink sector in the UK.

Procurement strategies are always drawn up in consultation with and apply across and throughout all departments and functions in a company, to ensure the identification, selection and use of the 'right' suppliers and resources that will enable a business to maximise profit and gain competitive advantage. The right suppliers are often considered to be those that can deliver at the right time and the right price and this has led some to question the nature of buyer–supplier relations (Badenhorst, 1994). For example, larger buyers have more purchasing power in the market place than smaller buyers and appear to operate competitively with their customers (Saunders, 1997). In contrast, small firms (SMEs) are more flexible and relatively more open with their suppliers because they have little purchasing power (Quayle, 2002). In the UK food and drink industry, for example, this has led to criticism of food processors' and multiple grocery retailers' treatment of farmers as suppliers (Commins, 1990; Welsh, 1997).

As a result of the need for suppliers to meet buyers', quite often highly specific, input requirements, there has been some reorganisation of buyer–seller relationships, into more collaborative or partner type approaches that bring suppliers, procurers and distributors together (Saunders, 1997). This approach to procurement is founded upon more direct, personal and close-knit commercial transactions between buyers and sellers involving knowledge transfer and exchange (Ghingold & Johnson, 1997; De Toni & Massimbeni, 1999; Nesheim, 2001; Humphreys *et al.*, 2002). Indeed, procurement representatives work closely with their suppliers from a very early stage in the product development process to ensure the ready and guaranteed availability of the goods and services they require. This approach to procurement and particular styles of management that have been linked with it are described next.

Management of procurement in the early 21st century food and drink industry

Changes in the macro economy, market maturity and consumer demand have stimulated a cultural shift in approaches to procurement. Early supplier involvement in the supply/production chain, the development of relationships with suppliers, the monitoring and assessment of suppliers on a continual basis, supplier certification and measurement of performance have become key principles in strategic, market responsive, procurement activities. Figure 4.1 illustrates how contemporary food consumer demands have shaped approaches to strategic food procurement and supply chain management principles and styles for particular strategic outcomes:

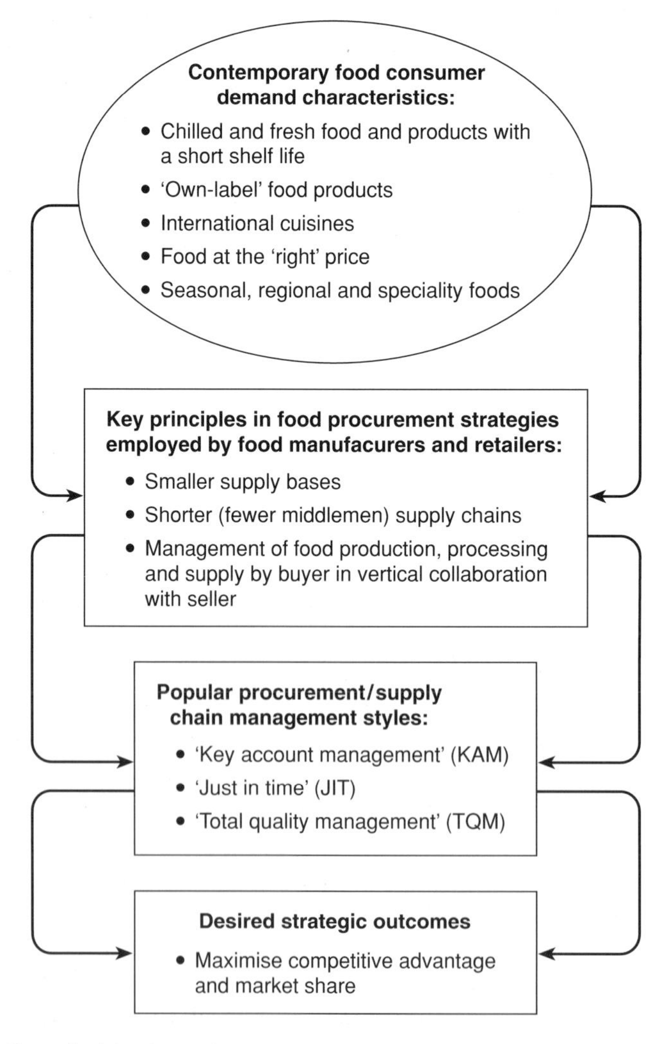

Figure 4.1 Key principles in contemporary procurement management.

these are now described and briefly illustrated in relation to the food and drink sector.

Consumer demand focused approaches to procurement and supply chain management

Key account management (KAM)

The KAM approach to procurement is developed upon an assessment of the true cost (all qualitative and quantitative transaction costs) and profit associated with a transaction between a buyer and seller (Steele & Court, 1996). The decision to pursue a KAM approach with a customer is dependent on each individual company's strategic business objectives, the relative attractiveness of the account and the competitive position of suppliers to the buyer, and vice versa (Steele & Court, 1996). Such approaches are time and human resource intensive and tend to evolve over time.

An ideal KAM model would be synergistic. Steele and Court (1996) suggest that the KAM approach engenders the following:

- more flexible approaches and efficient responses to consumer demand;
- lower stock levels and administration costs, leading to improved cashflow;
- improved long term planning of required volumes of inputs due to the availability and exchange of supply and demand data between the supplier and the procurement agent;
- innovation and technical advancement through better management and exchange of information throughout the supply chain;
- fewer, or eliminated, shortages in key resources;
- the ability to buy new resources for development into finished products in shorter times than competitors.

It has been suggested that partnership approaches to raw materials' procurement are the panacea for all possible supply chain problems (Steele & Court, 1996); for example, when suppliers are highly competitive and the materials that they provide are highly attractive to a procurement agent, then a collaborative relationship should be developed between the supplier and the procurement agent and should be defended to maintain the supply linkage. Furthermore, even when suppliers do not hold a competitive position in the market place and their resources are not particularly attractive to the procurement agent, it has been suggested that a procurement agent should withdraw from collaborative supply linkages on a selective basis, to retain the collaborative elements that add value to their performance (Steele & Court, 1996). In reality though, buyers and sellers generally relate to one another via one-off or short-term, quite adversarial interactions and transactions (MacDonald & Rogers, 1998), perhaps because of the need for proper regulation and monitoring of partnerships to ensure their continuity; this can be difficult to develop and maintain since, when relationships become closer, problems associated with conflict become more readily apparent (Davies, 1993).

Just in time (JIT)

Partnerships are a key element of the JIT approach to procurement which is perhaps most readily related to the Japanese car manufacturing industry. For example, Nissan have developed procurement linkages with suppliers in close geographical proximity to their manufacturing plants to ensure the delivery of parts and resources as and when required at the manufacturing site. This ensures low levels of, or no, stock holding at the manufacturing site and requires the supplier to be responsible for it for longer than under more standard procurement models where the key principle is to ensure supplies at least cost. In this way, the buyer (Nissan) has reduced raw materials' storage overheads and can use what would be storage space at their manufacturing site for more productive purposes.

The JIT approach is dependent on leadership and a clear communication of demands and objectives, requiring teamwork and collaboration internally and with suppliers; it has been successfully adopted in a number of economic sectors. However, it is reasonable to suggest that it may have limited application opportunities in the food and drink sector where the primary production of ingredients in a local area can be restricted due to the physical geography and natural local environment. That said, it was reported (IGD, 2002) that 85% of businesses, in the food and drink industry that have adopted a JIT approach have experienced significantly increased efficiency.

In the UK, JIT approaches to food and drink procurement have been referred to as 'efficient replenishment' techniques (Whiteoak, 1998). Efficient replenishment systems and techniques have been developed using the data collected by multiple and smaller grocery retailers when recording the barcodes of the produce they sell. The content and value of each individual grocery basket are recorded; the data are stored electronically and used by retailers to compile their stock replenishment (and therefore food and drink procurement) requirements. Retailers then inform their suppliers (food manufacturers) what they require, where and when. Two of the most important efficient replenishment techniques used by multiple grocery retailers in the UK are summarised in Box 4.2.

Whiteoak (1998) notes that these techniques (and others such as integrated suppliers, synchronised production and automated store ordering) can benefit the retailer enormously since they reduce the amount of working capital that is tied up in stock awaiting sale. In addition, they substantially reduce the human and capital resource requirements for warehousing. There has been a suggestion, however, that these techniques imbalance the retailer–manufacturer relationship as the retailer has control over the entire food and drink procurement process and, therefore, the actions of the food and drink manufacturers and producers. The manufacturer does, however, benefit from knowledge about market demand and access to the end market, and in theory, therefore, can co-ordinate and plan the procurement of raw materials for food and drink production and manufacture at least as and when required.

Box 4.2 Summary of the most important efficient replenishment techniques in the UK food and drink industry (after Whiteoak, 1998).

Continuous replenishment is a 'vendor managed inventory' approach to food and drink procurement. It can take a number of forms, but the most common take place in a 'fixed order quantity' environment where larger retailers review sales daily and use those data to forecast their future stock requirements from their suppliers, informing them of their demands through exchange of electronically stored data from product barcodes (using electronic point of sale (EPOS) systems). Manufacturers provide the required produce for the retailer by delivering to the retailers' regional distribution centre from which produce is transported to individual retail outlets. The retailer has ultimate control over the entire procurement process in this scenario. Another form of this approach is the 'fixed order cycle' used by smaller retailers less able to exert the same amount of control over their stock procurement practices. In this situation, it is more cost effective for smaller retailers to pre-arrange deliveries of particular stock at certain times, e.g. once every fortnight. Because of the amount of market share held by multiple retailers in the UK, this approach is used less and less.

Cross docking occurs when vehicles used to make deliveries to a multiple grocery retailer's regional distribution centre are, after having been emptied of the stock they arrived with, restocked with goods that they then transport to retail outlets. This technique is a traditional approach for the replenishment of perishable food and drink or that which has a very short life cycle.

Total quality management (TQM)

Another contemporary approach to food and drink procurement is the TQM approach which was developed around leadership at senior management level filtering through each department and function in a business to ensure quality maximisation to meet consumer and market demands (Stanley & Wisner, 2002). It is a key concept at the foundation of many industrial production activities and, in view of a high number of food scares in the UK, for example, has become crucial in contemporary, consumer and market demand responsive food and drink industry.

With regard to the issue of food safety in particular, it is notable that most UK consumers are extremely aware of the concepts of food safety and quality (Lacey, 1992; Dnes, 1996; Mintel, 1997; Almås, 1999; IGD, 1999; Shaw, 1999) as a result of a series of high profile food safety crises in recent years (Howells & Bradgate, 1990; Harrison *et al.*, 1997), so much so that despite the Food Safety Act being passed in the UK in 1990 (Simpson, 1992; James, 1997), the general public no longer have faith in the government to assure them that the food they eat is safe and high quality. Therefore a number of private sector food safety regulation systems have also been developed, including, for example, hazard analysis and critical control point (HACCP) analysis. HACCP systems relate to each stage of the food production and processing system, identifying and implementing the means to prevent hazardous food production (Loken, 1995). The systems are administered by local government Environmental Health Officers (EHOs) who assign a measure of risk to food companies on the basis of what food they use, how profitable their business is, and how much control they are able to exercise over the premises where food is produced (Harrison *et al.*, 1997; Simpson, 1992).

HACCP and other forms of private sector food safety management systems were widely adopted by multiple grocery retailers and larger food processors in the 1990s (Loken, 1995; Marsden *et al.*, 1998; Beardsell & Dale, 1999). Food quality regulation has therefore become a significant, yet unstable (Harrison *et al.*, 1997) vehicle of communication and authority between food businesses (Marsden *et al.*, 1998). Through their highly individualistic procurement strategies, the private sector in the food and drink industry is responsible for more quality regulation than any other (North, 1994).

Examples of the form and structure of consumer demand focused procurement relationships

Consumer demand responsive approaches to procurement have increased buyers' dependence on not only quality goods and services, but also reliable suppliers in all departments and across all functions in a business organisation (Steele & Court, 1996; Saunders, 1997). The implementation of this approach in many cases has involved the development of fewer, leaner, shorter and more direct and collaborative buyer–seller relationships. Some authors have referred to such a reorganisation of procurement relationships as the 'Japanisation' of western business practices (Steele & Court, 1996; Saunders, 1997) as these have sought to establish closer and more collaborative, less adversarial, linkages with their suppliers.

In further contrast to more conventional, arm's length, adversarial, supply linkages, more technologically driven and underpinned approaches to buyer–seller collaboration are gaining increased support. Electronic commerce (EC) is an effective business–business tool that has grown in popularity since the mid–late 1990s. It became an increasingly important medium for business–consumer relations via the Internet in the late 1990s (Mai & Ness, 1998). In its most basic form, EC exists as an Internet based portal that enables buyers and sellers to enter into supply linkages with one another using a common technology as a medium. Additional services can also be offered via such a portal; for example, logistics and industry news circulation (Varon, 2001). More interactive modes of exchange are also possible. For example, collaborative EC is a way in which companies can share intimate details about themselves and their businesses with the hope of gaining quicker access to new markets, reducing manufacturing time, maintaining low inventory costs, and responding to consumer demand more quickly (Varon, 2001). More specifically, Henderson (2002) has suggested that EC is most useful when commodities: (a) are easy to define and catalogue, (b) are frequently bought and the expenditure per transaction is relatively low, (c) are controlled by the purchasing function, (d) are used by different departments in the same enterprise, and (e) are provided by suppliers who are open to the idea of electronic selling. Here, commodities are defined as raw materials for food production and processing into finished, packaged products available on the retail shelf.

It should be noted that EC can incur substantial costs because each business requires the same IT system and operational skills to manage the data exchange, which is likely to be in stark contrast to more traditional paper and fax based purchasing systems. In addition, a number of specific costs/barriers to the use of EC have been identified (CIPS, 2001): data protection, payment security, contract formation, domain name allocation (and recall) and signature authentication.

Electronic commerce is therefore not always the most appropriate system for procurement and there is no single model/approach to the use of EC. Therefore, it is unlikely to be used by all business organisations and probable that individual businesses will choose the model that best suits them.

Developing a consumer demand oriented procurement relationship between buyers and sellers

The following buyer and seller characteristics are essential in more partnership oriented, contemporary procurement relationships that aim to meet consumer/market demand:

- a commitment to a long term supply linkage and trustworthiness;
- an ability to work with mutually agreed objectives and share risks and rewards amongst partners;
- partners being capable and competitive.

Such characteristics are not always noticeable or easily identified; therefore strategic procurement partnerships can take time to evolve (Steward, 1997; MacDonald & Rogers, 1998). They begin with a pre-relationship stage where the buyer and seller become aware of each other, then move on to the early stage when the buyer and the seller gain and exchange knowledge about each other; the development stage is next when the buyer and the seller begin to trade with each other; then comes the long term stage when the buyer and seller become embedded in a procurement relationship with one another. Finally, there is the final stage where the partnership may close, be renewed or continue.

The partnership approach can therefore help resolve a number of issues in contemporary procurement systems and relationships. However, partnerships are neither suitable nor easy to create in all cases (Steele & Court, 1996). Furthermore, due to the concentrated nature of collaborative relationships between individual suppliers and buyers, conflict is easily detected and may be difficult to overcome, thereby challenging the longevity of the linkage.

At this point, it is also noteworthy that such relationship forms and structures are rather uncommon in northern Europe or the developed western world, and in some cases have been implemented somewhat unwillingly as a means to an end. In the UK, for example, prime beef cattle suppliers in Northumberland have come together as a group of producers who market their produce under the label of 'North Country Primestock' as a means of resisting moves towards vertical integration and collaboration with larger food processors and the multiple grocery retailers as a result of their concern about a loss of direct self control over their own activities. This is because of the multiple grocery retailers' purchasing power and ability to encourage smaller suppliers to provide what they desire. For example, suppliers who want to sell their own produce to multiple grocery retailers with a larger share of the market are more likely to be asked to provide it to the retailer at low cost (Davies, 1993). Thus, the relative size of the buyers and sellers in a relationship affects participants' willingness to be involved in a partnership-based procurement process.

In addition, because different supply linkages have different strategic importance to respective buyers and sellers, it may be neither appropriate nor necessary to develop a long-term partnership with each and every buyer and seller. This is particularly the case if identical goods/services, i.e. of the same quality, volume, type and availability, can be procured elsewhere quite easily and at an appropriate price (Steward, 1997).

Case study of contemporary procurement management in the food and drink industry

At the end of the 20th century, Tesco, the largest multiple grocery retailer in the UK, pursued a number of approaches to more collaborative procurement, involving not only their suppliers at primary producer level (farmers), but also food processors and research and development institutions, one example of their partnership approach to food procurement being their 'Best Beef' scheme in which they aimed to increase the value of the beef chain from the calf to the finished produce on retail shelves (Tesco, 2002). (For this purpose, value is defined as beef quality, animal welfare, stock and meat traceability, hygiene management during production, processing and transportation, and overall assurance of that to the consumer.) The scheme was developed to address the problem of reduced beef quality and to reduce the number of surplus calves that are generally sold at a less than premium price. It is based on a partnership between Tesco, Express Milk Partnership (the dairy producer club that supplies them), Quality Calves (a farmer owned co-operative), calf rearers/finishers, dairy farmers, Southern Counties Fresh Foods, and the Meat and Livestock Commission (MLC) in the UK. The scheme involves the production of calves by dairy farmers using bulls from the Best Beef Sire list held by the UK Meat and Livestock Commission. Calves can then be finished by individual farmers or sold to the Southern Counties producer club members through Quality Calves and then to Tesco; the entire production cycle is shorter, providing what Tesco claims to be animals of better eating quality with a fully traceable and assured production process and history.

Another initiative funded by Tesco is the Tesco Centre for Organic Agriculture: this is a collaboration between Tesco and the University of Newcastle upon Tyne, UK, to carry out research into the development of organic food. The Centre compares conventional and organic production methods to provide scientifically sound assessments of their impact on health, food quality and the environment. Findings from the work will ultimately be used in Tesco's organic procurement strategies and processes to help increase their organic range and consumer demand for organic foods.

Summary

The act of procurement has evolved from a standardised, administrative task in the mid-20th century into a strategic function that directs all activities in, and operates at the heart of, a business organisation in the early 21st century. Partnerships between buyers and sellers, as opposed to more adversarial and competitive relationships between buyers and sellers where the market and consumer demand is of secondary importance to them, offer profit and market share maximisation opportunities for those involved in the procurement process and the entire supply chain. Such approaches do, however, require a substantial cultural shift in business organisation and practice, but have been successfully implemented in the UK food and drink industry by the UK's largest multiple grocery retailer (Tesco). In the food industry in particular, consumer demand for safety and quality assured food that can be traced from farm to fork is encouraging closer procurement relationships. On balance though, in the future, MacDonald and Rogers (1998) suggest that it is likely that two types of procurement will evolve as the norm. Global companies are more likely to pursue leaner, more synergistic and integrated supply chains, particularly when manufacturers are involved, whilst smaller suppliers are more likely to engage with local buyers and involve non-critical suppliers and virtual, if not actual, acquisition of the supplier company.

Study questions

1. Describe the changing role of procurement in supply chain management between the mid 20th and early 21st century.
2. Describe the core principles at the heart of more contemporary approaches to procurement management and briefly outline the form of a relationship that enables these approaches to be used.
3. Using an example from the industry, describe how and why the food and drink industry has developed and adapted to more contemporary approaches to managing strategic procurement and critically evaluate the value of this to food consumers.

References

Almås, R. (1999) Food trust, ethics and safety in risk society. *Sociological Research Online* **4**(1) (http://www.socresonline.org.uk/4/3/almas.html).

Atkins, P. & Bowler, I. (2001) *Food in Society. Economy, Culture, Geography*. Arnold, London.

Badenhorst, J.A. (1994) Unethical behaviour in procurement: a perspective on causes and solutions. *Journal of Business Ethics* **13**(9), 739–745.

Beardsell, M.L. & Dale, B.G. (1999) The relevance of total quality management in the food supply and distribution industry: a case study. *British Food Journal* **101**(3), 190–201.

CIPS (2001) E-purchasing cycle. Chartered Institute of Purchasing and Supply, London. (http://www.cips.org/technical/downloads/ecyclePDF).

Commins, P. (1990) Restructuring agriculture in advanced societies: transformation, crisis and responses. In *Rural Restructuring* (T. Marsden, P. Lowe & S. Whatmore, eds) pp. 45–76, David Fulton, London.

Davies, G. (1993) *Trade Marketing Strategy*. Paul Chapman, London.

De Toni, A. & Massimbeni, G. (1999) Buyer–supplier operational practices, sourcing policies and plant performances: result of empirical research. *International Journal of Production Research* **37**(3), 597–619.

Dnes, A. (1996) An economic analysis of the BSE scare. *Scottish Economic Society* **43**(3), 343–348.

Ghingold, M. & Johnson, B. (1997) Technical knowledge as value added in business markets – implications for procurement and marketing. *Industrial Marketing Management* **26**(3), 271–280.

Harrison, M., Flynn, A. & Marsden, T.K. (1997) Contested regulatory practice and the implementation of food policy: exploring the local and national interface. *Transactions of the Institute of British Geographers* **NS22**, 473–487.

Henderson, A. (2002) *Good Practices in eProcurement*, Chartered Institute of Purchasing and Supply, London.

Howells, G. & Bradgate, R. (1990) *Blackstones Guide to the Food Safety Act 1990*. Blackstones, London.

Humphreys, P., McIvor, R. & Huang, G. (2002) An expert system for evaluating the make or buy decision. *Computers and Industrial Engineering* **42**(2–4), 567–585.

IGD (1999) *Conference on Consumer Trust and Genetically Modified Food*. Institute of Grocery Distribution, Watford.

IGD (2002) *Process Improvement Demystified*. Institute of Grocery Distribution, Watford (http://www.igd.com/).

James, P. (1997) *Food Standards Agency: An Interim Proposal*. Rowett Research Institute, Aberdeen.

Lacey, R. (1992) *Unfit for Human Consumption*. Grafton, London.

Loken, J.K. (1995) *The HACCP Food Safety Manual*. John Wiley & Sons, New York.

MacDonald, M. & Rogers, B. (1998) *Key Account Management. Learning from Supplier Perspectives*. Chartered Institute of Marketing with Butterworth Heinemann, Oxford.

Mai, L.-W. & Ness, M. (1998) Perceived benefits of mail order speciality foods. *British Food Journal* **100**(1), 10–17.

Marsden, T., Harrison, M. & Flynn, A. (1998) Creating competitive space: exploring the social and political maintenance of retail power. *Environment and Planning* A **30**, 481–498.

Mintel (1997) *Food Safety*, Mintel Marketing Intelligence. Mintel, London (http://www.mintel.com).

Nellore, R. & Soderquist, P. (2000) Portfolio approaches to procurement – analysing the missing link to specifications. *Long Range Planning* **33**(2), 245–267.

Nesheim, T. (2001) Externalization of the core: antecedents of collaborative relationships with suppliers. *European Journal of Purchasing and Supply Management* **7**(4), 217–225.

North, R. (1994) The prevention of food poisoning: a strategy for deregulation. *British Food Journal* **96**(1), 29–36.

Quayle, M. (2002) Purchasing in small firms. *European Journal of Purchasing and Supply Management* **8**(3), 151–159.

Saunders, M. (1997) *Strategic Purchasing and Supply Chain Management*. Chartered Institute of Purchasing and Supply with Pitman Publishing, London.

Shaw, A. (1999) What are 'they' doing to our food? Public concern about food in the UK. *Sociological Research Online* **4**(3) (http://www.socresonline.org.uk/4/3/shaw.html).

Simpson, S. (1992) The Food Safety Act: food manufacturer response and attitude. *British Food Journal* **94**(1), 3–7.

Stanley, L. & Wisner, J. (2002) The determinants of service quality: issues for purchasing. *European Journal of Purchasing and Supply Management* **8**(2), 97–109.

Steele, P.T. & Court, B.H. (1996) *Profitable Purchasing Strategies. A Manager's Guide for Improving Organisational Competitiveness through the Skills of Purchasing*. McGraw Hill, London.

Steward, C. (1997) *Managing Major Accounts. Shaping and Exploiting your Firm's Intangible Assets*. Marketing Society with McGraw Hill, London.

Tansey, G. (1994) Food policy in a changing food system. *British Food Journal* **96**(8), 4–12.

Tesco (2002) The Best Beef Scheme: increasing the value of the beef chain from the calf to the counter. (http://www.tescofarming.com/pdf/bestbeef.pdf)

Varon, E. (2001) *The ABCs of B2B*. E-Business Research Center, Boston, MA (http://www.cio.com/research/ec/edit/b2babc.html).

Virolainen, V.M. (1998) A survey of procurement strategy development in industrial companies. *International Journal of Production Economics* **56/57**, 677–688.

Warde, A. (1997) *Consumption, Food and Taste: Culinary Antinomies and Commodity Culture*. Sage, London.

Welsh, R. (1997) Vertical co-ordination, producer response, and the locus of control over agricultural production decisions. *Rural Sociology* **62**(4), 491–507.

Whiteoak, P. (1998) Rethinking efficient replenishment in the grocery sector. In *Logistics and Retail Management. Insights into Current Practice and Trends from Leading Experts* (Fernie, J. & Sparks, L., eds) pp. 110–140. Kogan Page, London.

Chapter 5

The UK Livestock System

D. Harvey

Objectives

(1) To provide sufficient information about the UK livestock supply or marketing chains to allow readers to appreciate their specific characters and circumstances.
(2) To analyse the consequences of these characteristics for the development and management of the chains.

Introduction

This chapter outlines the major features of the UK livestock supply chains, beginning with a historical overview of the UK livestock sector and then turning to a consideration of the character of the three major livestock supply chains (dairy, intensive white meat livestock and red meats). It concludes by examining the red meat sector and supply chain in more detail, including reference to its competitive performance.

Historical overview

The overall post-war history of the UK livestock and livestock product markets is illustrated in Figure 5.1. In the UK we eat only half the number of eggs and drink one-third of the volume of milk that were consumed in the 1960s. While our total meat consumption has remained relatively constant, the composition has altered dramatically (Figure 5.2). Red meat (beef and lamb) consumption has fallen by three-fifths since its peak in the 1950s, pork consumption has remained relatively static, while poultry consumption has increased dramatically from virtually nothing in the middle of the last century, to become the clear meat of choice at the turn of this century. Of course, we also eat far more meat and livestock products in processed form, including meals eaten out rather than in the home. The catering trade now accounts for about one-third of total food consumption. Total spending on carcass meats has shown a substantial decline over the last 25 years, especially

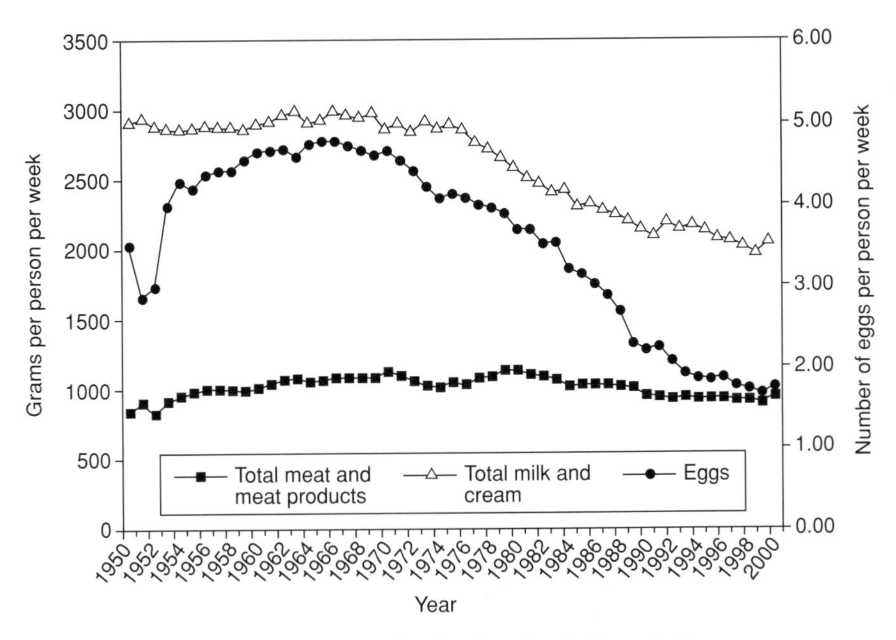

Figure 5.1 Weekly per capita consumption levels – livestock products (Source: DEFRA, 2002a).

Figure 5.2 Weekly per capita consumption of meats (Source: DEFRA, 2002a).

for carcass beef and veal, although spending on other meats (poultry and prepared and processed meats) shows a modest increase.

In common with most of the western developed world, spending on meat consumption in the UK has been falling in real terms even faster than the decline in total food spending. Between 1981 and 1992, total meat spending fell as a fraction of total food spending from 31% to 26%, while spending on carcass meats fell from over 14% to just under 9% of total food spending. More recent data in the UK have been confounded by a succession of food health scares, notably *Salmonella* in poultry and eggs, and BSE in beef, while the foot and mouth disease (FMD) epidemic in 2001/2 has further confused and distorted the longer term trends. Nevertheless, this is a picture of a generally declining industry. It is well known that the proportion of consumer spending devoted to food falls in real terms as incomes increase. Because consumer spending on food ultimately generates the incomes of all the people involved in the food chain, this means that fewer people are able to earn a full time living from the livestock supply chain as economic growth occurs and incomes improve. Livestock supply chains will, as a consequence, be under considerable economic pressure.

The character of livestock supply chains

From a marketing or supply chain perspective, the UK livestock food chain, like the systems of most other developed countries, separates into three rather different sectors: dairy and dairy products, white meats (pigs and poultry) and red meats (beef, mutton and lamb). All these supply chains have their roots in farms, which are generally atomistic (very many relatively small-scale fabrication plants), widely dispersed and frequently remote from the final consumer. All deal in highly perishable, fragile and vulnerable materials, which need careful handling and processing throughout the chains to guarantee quality and safety of the final product to the consumer.

Traditionally, these many farms were connected to final consumers through a chain of local livestock markets and dealers, to small-scale local processors, delivering to or selling through wholesale markets (of which Smithfield Meat Market in London is the most well known) to retailers and the occasional caterer. Milk, however, has always been somewhat differently traded, primarily because of its extremely short shelf-life and vulnerability to deterioration. Primary product auction markets, dealers and wholesale markets played no role in the supply chain for milk and dairy products. Rather, local dairies purchased supplies directly from the farm (before 1992 through the UK Milk Marketing Board as the farmer-owned monopoly seller of farm milk) for both delivery of fluid milk and processing of dairy products. Milk is delivered directly to the retailer, traditionally via the door-step delivery system and, more recently, to the major supermarket chains which now account for three-quarters of fluid milk sales (approximately half the total use of raw milk in the UK).

The major differences between the three sectors are summarised in Table 5.1. Given the limited space available here, this chapter will concentrate on the most complex and interesting of these supply chains, the red meats. However, before turning to consideration of the detail of the red meat supply chain, it is appropriate to outline the major features of each of the livestock chains.

Table 5.1 Major features of UK livestock supply chains.

	Dairy	White meats	Red meats
Primary production	Specialist, intensive Production plant: concentrated Fabrication time: daily Primary product: homogenous	Specialist, intensive Production plant: concentrated Fabrication time: weeks Primary product: homogenous	Generalised, extensive Production plant: dispersed Fabrication time: months/years Primary product: variable and heterogenous
Primary processing	Concentrated and simple	Concentrated and complex	Dispersed and complex
Secondary processing and distribution	Limited and concentrated	Considerable but concentrated	Considerable and dispersed

Geographical distribution and business structures of livestock farms

In the UK, dairy farming is primarily located in the wetter, western lowland areas of the country (south-west peninsula, Lancashire, Cheshire, etc.) where high quality and efficient grassland production is possible, although there are significant areas of dairy farming elsewhere. Once established and appropriately maintained, with new, younger cows to replace those at the end of their productive life (7 years or so), a dairy herd is a more or less continuous production system, providing milk on a daily basis for 270 days per cow per year. The capital requirements for a dairy herd are made up of both the considerable investment in the herd itself, the land needed to provide the basic feed (especially grass and other fodder crops in the UK) and the specialist milking and feeding plant and equipment. This tends to lead to dairy farms, or dairy units within the larger farms, being largely independent and specialist units.

White meat and egg production are predominantly based on grain as a feed and are typically carried out in large-scale units of environmentally controlled livestock housing (intensive systems) whose location is usually close to grain supplies (the eastern part of the UK). More recently, however, more extensive, free-range and organic systems have become viable, provided that their products can be distinctively marketed at a sufficient premium to cover additional production costs.

Pigs and especially poultry grow to finished weights and conformations quickly, in a matter of weeks. Furthermore, the fecundity of the animals means that the breeding herds and flocks necessary to supply the fatstock (or laying flocks) are much smaller as a proportion of the total population than is the case with cattle and sheep. Consequently, the capital requirements for these production systems tend to be dominated by housing and feeding equipment and the growing animals themselves, and are generally lower per unit of product than in the other sectors. In particular, since these animals cannot be produced and maintained on grass as they are not ruminants, the specific land requirements for these units are minimal, lending themselves to factory farm production. Per unit of protein, these enterprises are capable of producing much cheaper food than either the dairy or red meat sectors. In addition, the fecundity of the breeding herds and flocks allows for rapid development of the most efficient breeds which, coupled with controlled environment feeding and production systems, has produced an impressive improvement in technical efficiency in this sector of livestock production over the last half century, allowing it to become highly specialised and concentrated in fewer farming units compared with the dairy and red meats sectors.

Beef and sheep farms, as with dairy farming, are dependent on good supplies of grazing land, though are better suited, especially for the breeding herds and flocks, to using more marginal and upland areas since they do not need to be milked or cared for. Hence, they tend to be situated in the west and north in hillier areas of the UK. The breeding herds and flocks are the predominant part of the capital investment required, while the working capital required is also tied up in the animals themselves, which take months to years to mature to their finished states (grass-fed beef can take more than 30 months to fatten, though purely grain fed animals can be finished within 12 months). Not only do beef and sheep farms occupy the poorer agricultural land, they also typically provide poorer returns to their owners and operators (Figure 5.3). With the exception of pigs and poultry, these rates of return are considerably lower than would be regarded as commercial in most other areas of the economy. Returns to upland (less favoured area) farms are typically somewhat higher than their lowland counterparts, because of

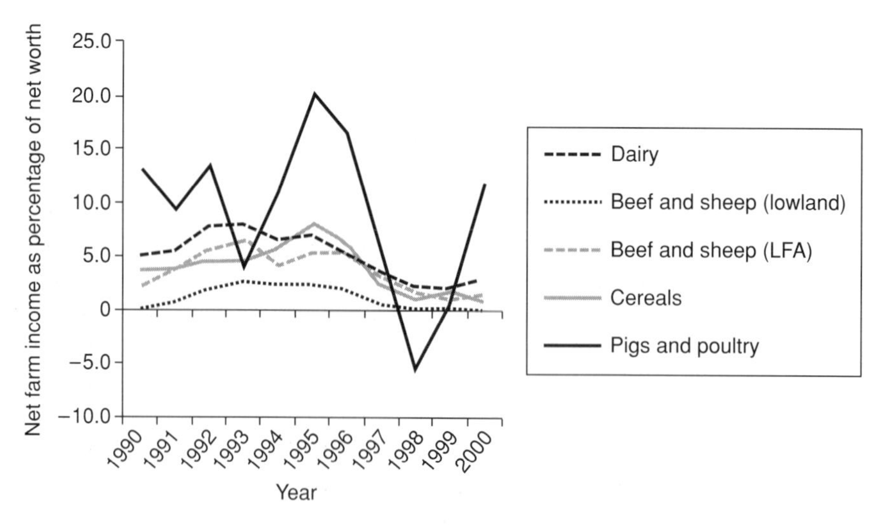

Figure 5.3 Farm rates of return, by farm type in England (1990–2000). These rates of return are defined as net farm income divided by the opening valuation net worth for each farm type; LFA means less favoured areas (hill farms) (Source: DEFRA, 2002b).

favourable support payments to these areas. However, higher support payments will become capitalised in the asset values for these farms, and thus will not affect the rate of return in the longer run. The relatively lower rates of return on lowland farms are typically offset by other (frequently cereal) enterprises, using otherwise surplus flows of labour, management, capital and land not needed or useful for cereal production. In addition, low rates of return are frequently said to reflect the willingness of farmers to count their independence and way of life as part of their return, which is clearly less important for factory-type pig and poultry units.

Some 50–60% of total beef supplies in the UK originate from the dairy herd, in the form of redundant dairy cows and surplus dairy calves (a dairy cow has to produce a calf before she will begin her annual lactation cycle). But there is a distinct trade-off between breeding animals that are good at producing milk and those which are good at growing fast and producing good quality meat. Dairy cattle do not, generally, produce meat of the highest quality nor are they the most efficient meat-producing animals.

The introduction and development of artificial insemination (AI) of dairy cattle has made a substantial difference to the organisation and management of dairy herds. AI allows the herdsman to achieve well-timed conception more accurately, ensuring that dairy cows return to their lactation cycles more efficiently. Furthermore, it allows the use of a variety of bulls, tailored to the qualities of a particular cow. The best milkers are used to produce replacement cows, by being bred to high performance dairy bulls, while those cows with less good milking performance are crossed with high performance beef bulls so that the calves will be more suited to meat production. Eliminating the need to manage dairy bulls that are aggressive and difficult to manage within the productive dairy herd is an added and significant advantage of AI.

The consequence of this has been that the quality (feed conversion efficiency and milk quality) of dairy herds has been very substantially improved over the past half century, with dairy herd efficiency improving at 2–3% per year. Dairy production necessarily involves close and daily attention by the herdsman to the cows, which means that AI is a relatively straightforward, and thus cheap, technology to employ in the dairy herd, unlike a pure beef or sheep enterprise. Here, AI has yet to become a common practice because of the substantial additional work and effort required in these enterprises. As a result, technical improvement in beef and sheep production has been more difficult to achieve, and costs of primary production have neither been reduced nor made less sensitive to varying conditions of both nature and nurture in the red meat sector, unlike the other two livestock sectors. Nevertheless, the dependence of the dairy herd on twice (sometimes more frequent) milking and on grazing land has restricted the size of individual enterprises to herds that can be efficiently managed as a single unit (currently about 300 cows or so, though there are examples of 1000 cow units which are mainly fed on purchased feeds).

A little more detail on the structure of the primary production end of the livestock supply chains is shown in Table 5.2. The concentration of total production among primary producers is very much higher in the poultry and pig sectors than

Table 5.2 UK livestock primary production structures (DEFRA, 2000).

	Total animals (million)	Total flocks/herds	Percentage of animals in commercial units[a]	Flocks/herds as commercial units[a] (%/number)
Dairy	2.3	32 000	74	44/14 000
Pigs	0.7	7 500	88	8/600
Poultry	105	2 000	97	50/1 000
Eggs	36	24 000	95	3.7/880
Sheep	19	78 000	82	36/28 000
Beef	1.8	65 000	62	22/14 000

[a] Commercial units are here defined in an arbitrary fashion (as allowed by the census classifications) as: (a) dairy herds (70 cows or more), (b) pig herds (100 breeding pigs or more), (c) poultry flocks (20 000 birds or more), (d) laying hens (5000 birds or more), (e) sheep flocks (200 ewes or more), (f) beef herds (40 cows or more).

in the dairy, beef and sheep sectors. The beef and sheep sectors, especially, show very considerable fragmentation with large numbers of relatively small production units.

The dairy supply chain

Dairy farms tend to be specialist units, concentrating on milk production (though many dairy units are also operated in conjunction with arable enterprises). In the interests of efficiency and competitiveness at the farm level, there has been a trend towards fewer, larger herds, each yielding more milk per year (now averaging almost 6000 litres per cow). Many of these dairy farms breed their own replacement dairy cows so that their production plant (the cow herd) is concentrated within a single business and location. Calves, a necessary by-product to milk production, that are surplus to replacement requirements are either kept and fed on for beef or sold to other beef or veal farmers. Recently, the French export market for live young dairy calves has been closed because of animal welfare and health concerns (BSE and then FMD). This market was valuable to UK producers, because the French market both values veal production more highly and is less restrictive in production methods than the UK. Closure of this market led to a substantial fall in dairy calf values in the UK, sufficient in some cases to make it uneconomic to try to sell these animals on for further feeding.

Dairy farms typically market their milk through medium-term contracts to dairy processors, with some 32 000 dairy herds in the UK supplying 135 buyers, of which only 21 buy almost 89% of the total UK milk output (Dairy Council, 2001). Half of the total milk produced is sold and consumed as liquid milk, with the other half being processed into dairy products (Figure 5.4).

Seventy per cent of household liquid milk purchases are made through supermarkets (Dairy Council, 2001), with an even larger percentage of other dairy products being purchased through these outlets. Total milk demand is relatively static to declining, with a decline in liquid milk and cream consumption offset by a

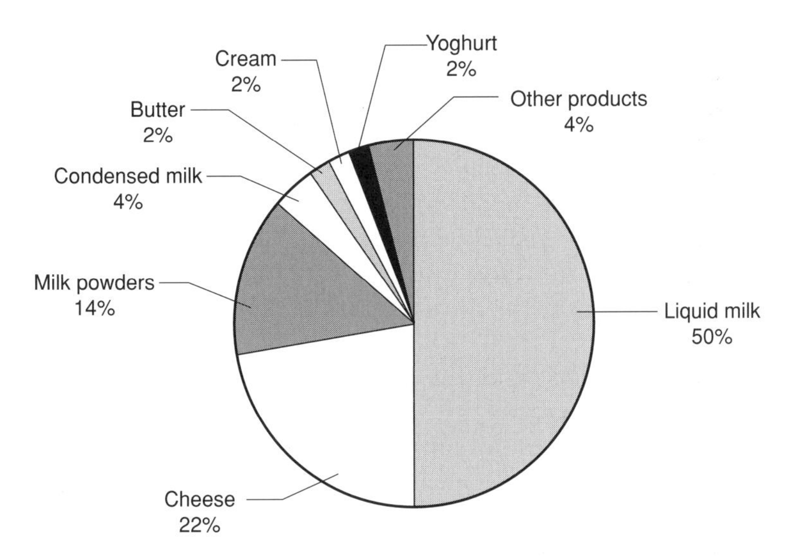

Figure 5.4 Utilisation of milk during 2000 by dairies in the UK (Source: Dairy Council, 2001).

modest increase in some other products, especially certain cheeses and fresh dairy products such as milk, butter and yoghurt.

Even more than other livestock sectors, the dairy sector is heavily regulated (see, for example, Dairy Council, 2001). Total deliveries from farms are restricted by delivery quotas under the dairy regime of the EU's Common Agricultural Policy, which has its own rather significant effects on the production structures, costs and marketing patterns (Colman, 2002). In addition, there are a number of health and safety regulations governing the production, processing, distribution and retailing of milk to ensure the quality and safety of the final products. Furthermore, the marketing arrangements for milk from the farms to the dairies has undergone a major change in the recent past, from being controlled by producer marketing boards to being deregulated, which has also resulted in some considerable restructuring of marketing channels and practices. DEFRA (2001a) has provided some further considerations pertaining to the UK dairy supply chain.

The white meat supply chain

Pigs, poultry and eggs tend to be produced intensively in controlled environment houses with large-scale herds and flocks, and relatively few production units. Of all the livestock sectors, this is the most industrial in nature. This production structure has developed because of the twin factors of consumer demands for uniform and cheap products and the technological possibilities available for the primary production animals. Animal breeding and feeding technologies have allowed the development of highly productive pigs and poultry with very high feed-conversion rates, especially if kept in controlled environment conditions. However, there is a small though growing market for more extensively produced products (organic pork, free range poultry and eggs, etc.) reflecting increasing concerns among at least some segments of the retail market for more animal and environmentally friendly production practices. More friendly production systems typically require larger land areas (free-range conditions), tend to be less technically productive and have higher costs.

The industrial nature of the major production systems is reflected in the organisation and operation of the supply chains. Much of the total production is highly vertically integrated, with production systems, feeding regimes, breeds and types of animal being controlled to a large extent by the processors, and serviced through rolling contracts or even being carried out as part of the same production and processing business.

The industrial nature of the intensive livestock supply chain is reflected in the processing of these products into varied and various final consumer lines. One consequence of this major processing activity is the extent to which the raw materials (pigmeat, poultrymeat and eggs) are considered relatively homogenous compared with beef and sheep. British supplies are very often in direct competition with sources from offshore. This is particularly well illustrated by the domination of offshore, especially Danish, supplies of bacon and ham to the UK market. Although pigs produce both bacon and ham as well as pork, both the production and especially the curing processes are different for each product. The Danes, especially, have tended to concentrate their production and processing capacity on the production of quality bacon and ham, much of which they export successfully, particularly to the UK, which is generally only about 50% self-sufficient in these products.

The extent to which offshore supplies replace domestic supplies as the source of choice, for either processors or retailers, depends to a considerable extent on relative prices at which commodities of broadly similar qualities are available, as well as on established supply chain networks and connections and marketing activities as in the Danish bacon case. There are two major drivers of relative prices aside from any government policy intervention through border protection, import taxes and export subsidies: the local costs of production and the exchange rate. It is noteworthy that despite the well-known existence of the Common Agricultural Policy, which typically does provide considerable border protection for EU supplies versus those of countries outside the EU, the extent of support provided for intensive livestock has actually been very small. This lack of protection (within a general historical climate of substantial protection and support) is for a very good reason: any slight advantage offered to domestic supplies can be expected to have a major effect on the size of the primary supply sector. Without any significant constraint on the resources (especially land) needed for these enterprises, and with no significant lag between decisions to expand production and the capacity to produce more, these sectors respond rapidly to price incentives, which, if set just too high, quickly result in unsaleable surpluses. As a result, the EU intensive livestock sectors have only been protected from international competition to the extent necessary to offset the effects of EU agricultural protection on the cost of feed (cereals).

These factors are well illustrated in Figure 5.5 (especially in the case of pork). So long as the sterling exchange rate remained relatively undervalued against other

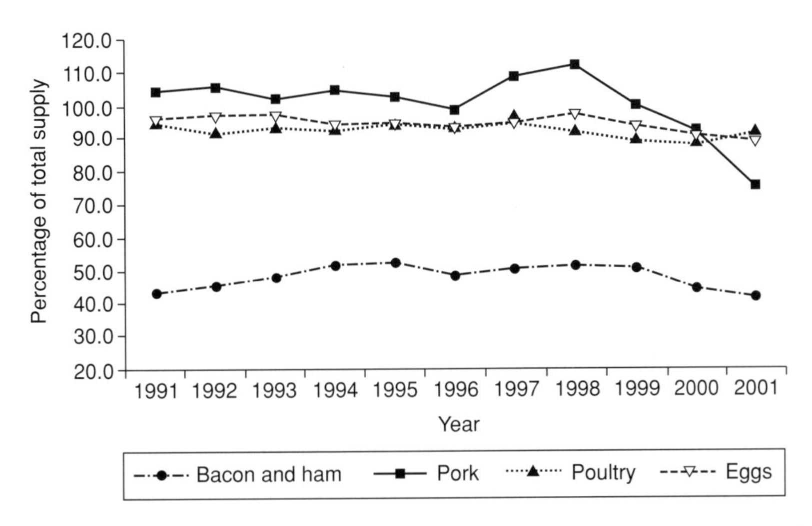

Figure 5.5 Domestic production as a percentage of total supplies (intensive livestock) (Source: DEFRA, 2002c).

currencies, especially in 1996/97, domestic UK pork production was very competitive with offshore supplies and the UK tended to be more than self-sufficient in pork. However, since 1997/98, sterling has been relatively strong, especially against the euro, and continental supplies have become significantly more competitive as a result. Furthermore, consumer and constituent pressures in the UK have been sufficient to encourage rather more restrictive regulation of sow and pig rearing practices in the interests of animal welfare. These more restrictive regulations have increased the costs, at least temporarily, of rearing pigs in the UK compared with continental competitors, resulting in a substantial fall in the national UK pig herd and a fall in the UK self-sufficiency rate.

It is ironic that a domestic UK concern over animal welfare, while leading to a substantial reduction in less acceptable practices in the UK, is leading to greater reliance on offshore supplies which do not necessarily respect these higher welfare standards. We might note, in passing, that it is one thing to regulate a local industry so that it follows expressed constituent concerns, and quite another to oblige those same constituents to bear the costs of imposing their concerns and preferences.

The red meat sector

UK market background

Traditionally, the lines of communication and information flows between the several links in the red meat supply/marketing chain were informal and relied upon an established trust between participants, born of long experience. Thus, hill farming areas were largely the source of lowland breeding livestock, whose purchasers regularly used the same autumn auction markets, often relying on the same farms for their replacement animals. Further down the chain, local butchers, often associated with local slaughterhouses, also relied on known markets and known farmers for their supplies, selling in turn to known families and households. The average movement of animals and carcasses was relatively low, with the exception of major movements of replacement breeding stock from the uplands of the north to the lowlands and downlands of the south. The economist's perfect market was supported by an intangible and informal web of networks and established reputations, which generated considerable trust and integrity in the system as a whole. The possible tendency for separate and somewhat isolated markets to diverge substantially from each other was kept in check by dealers, seeking to buy cheap and sell dear, known in economist's jargon as arbitrage. This ensured that both occasional surpluses in one market were quickly moved to deficit markets and that prices seldom differed by more than minimal amounts.

Developments in, especially, the retail part of the chain have led to the erosion of this traditional, fragmented but informally networked system. Changing shopping, food storage, preparation and eating habits, associated with the growth of the supermarket chains, led to concentration of the retail outlets and decline of the traditional local butcher. Major multiples gain competitive advantage over the traditional high street butcher through centralised buying and distribution, as well as through provision of one-stop shopping.

The centralisation of distribution is also associated with concentration of processing and the demand for mechanistic control over the processing and distribution chain, emphasising continuity of quality-controlled supplies. In the grain-fed livestock sectors (pigs and poultry), this has been substantially achieved through effective vertical integration of the chain from primary producer to retailer, albeit seldom under single firm control. With grazing livestock, however, the conditions of primary production proved too variable and fragmented for vertical integration or rigid supply contracts. Nevertheless, the major multiples are attempting to regulate and manage their supply chains through concentration on particular plants and particular sources, emphasising the importance of the buyer networks, the attraction of forward contracts, of electronic auction marts and, at least to some retailers, of dedicated suppliers and traceability.

Gunthorpe *et al.* (1995) summarise the state of the red meat sector at the turn of the 21st century as follows: 'The changing socio-economic environment as seen in the consumers' changing lifestyle, work and leisure patterns has resulted in a rapidly changing market for red meat. As a result of the severe competition in the food market, the share of red meat within the total meat market is under constant pressure' and identify three 'powerful structural themes' to the current development of the sector:

(1) The end market is steadily shifting away from the business of selling more traditional cuts of meat to consumers, towards the sale of more value-added meat and meat as a food product ingredient.

(2) Despite the increasing concentration of slaughtering and processing capacity, the industry will remain highly competitive (that is, plagued with excess and outdated capacity) with persistently narrow margins.
(3) The red meat industry is seeking to develop structures that will lead to more integrated production and marketing throughout the meat chain, in order to deliver the consistent quality of product required by the market.

The retail market

The UK (with Germany and Sweden) has amongst the lowest per capita consumption of beef and veal in the western world, with the exception of Japan. The Americans consume three times as much beef as the British, while the Italians and French consume 40% more per head. However, the British consume slightly more mutton and lamb than most other richer peoples, except the Australasians and also, importantly, the Middle Eastern peoples.

Many of the red meat categories were experiencing falls in consumption in the UK before the BSE crisis: beef and veal consumption fell by over 14%, bacon by 19%, and mutton and lamb by almost 20% between 1983 and 1993. Poultry (white meat), in contrast, saw strong growth of 40%, while pork consumption increased by 9% over the same period. The decline in demand for carcass beef in the UK has been even more pronounced, echoing similar, though less severe, erosion of demand in most advanced western societies. This declining trend is even evident in Australasia and North America, where beef-eating cultures have been more strongly established than in the UK. However, catering use of red meats has grown by 18% over the period 1983 to 1993 (in volume terms) against a fall of 15% at retail/household level (with an overall decline in volume of 8.4%) and a resultant growth in catering's share from one-fifth to over a quarter, with the real gains being for beef and pork, eaten as either fresh or processed products. The major UK supermarkets were handling 55% of the red meat, bacon and poultry retail household sales by 1997 (cf. 26% in 1979), while the independent butchers' share declined from 47% to 25% over the same period, with the balance accounted for by co-operatives, independent grocers, freezer centres, farm shops and market stalls. Farm shops and farmers' markets are a recent, visible, but still very small, counter to this general trend towards retail concentration.

While total consumer spending grew by 75% in the ten years from 1983 to 1993, food spending only increased by just over 13%, and meat spending increased a mere 5.5%. However, total spending on meat products rose by 14% in real terms over the period 1985–1993 and accounted for 35% of all meat spending by 1993, before BSE. This shift in consumption away from meat purchased as fresh or frozen cuts to meat products and meat-based convenience foods tallies with the general trend in lifestyle and consumer demand, whereby convenience and timeliness are an important part of food products. For meat suppliers, this means extra preparation and pre-cooking before the act of consumer purchase. Yet, as noted by the chief meat buyer for one of the major multiples at the Agra Europe Meat

Conference in October 1996, 'it is a sad comment that the only new beef product to hit the market successfully since the war is the hamburger'. This speaker also noted that his division showed the strongest turnover performance of the whole company over an unspecified period, indicating both intense competition at the retail end of the chain, and major efforts in some parts of this chain towards greater control and greater responsiveness to customer requirements and demands.

The Food Safety Act 1990 replaced and extended the provisions of the Food Act 1984 and placed particular emphasis on aspects of food safety. It introduced the defence of 'due diligence' into food law, prompting many companies to consider introducing a 'due diligence system', and has been a major reason why many large supermarkets have become concerned with traceability, and the introduction by many abattoirs of process control specifications, in the form of the Meat and Livestock Commission's (MLC) Blueprints for Eating Quality (Altenborough, 2003). Over 60% of the high value cuts of beef are now produced to the blueprint specification, which forms the basis of the high quality cuts specification in the major multiple retailers. There is increasing interest in the industry at all levels to develop a more co-ordinated market approach in the meat chain, to give consistency of quality supply, traceability and assurance for the consumer.

Producers are also responding to consumer concerns over animal welfare and the environment by providing guarantees through a variety of farm assurance schemes. The most notable of these in England and Wales are the Farm Assured British Beef and Lamb Scheme (FABBL), the British Quality Assured Pig Initiative and the Farm Assured Welsh Lamb. In Scotland, there is Scottish Farm Assured Beef and Lamb, and the Scottish Pig Initiative. Reflecting these trends, and reinforcing the motivations, a Gallup opinion survey in 1995 reported that 71% of people thought that retailers have a responsibility to animals, while 60% felt it important for the food industry to treat animals humanely. Meanwhile, the MLC has been responsible for a major advertising campaign to persuade consumers of both the health and safety of British meats and the ease with which these foods can be prepared and cooked (through apparently popular and widely recognised recipe cards).

The red meat policy context

In addition to the substantial changes in marketing structures and trends, the red meat sector also faces very considerable changes in its policy environment. The traditional support and protection of the market provided by the EU Common Agricultural Policy consisted of border protection. Taxes (import levies) were charged on all imports from countries outside the EU, to increase the price of these competitive supplies to the supported European level. In the event that the EU market developed surpluses of production over consumption, export subsidies were paid to allow these surpluses to be sold at world prices. Otherwise, surpluses accumulated and deteriorated in the EU's intervention stores at the taxpayers' expense. The European beef market, in particular, has been plagued by surpluses since the early 1980s, indicating (at least to an economist) that the domestic European price was too high. In fact, during the early 1990s, European beef prices were consistently some 2.5 times greater than their Australian and Argentinian counterparts, and 40% greater than the equivalent US prices. Although the EU has generally been a net importer of mutton and lamb, especially from New Zealand, a similar domestic premium in market prices has been maintained by border protection.

This traditional protection system is no longer sustainable. The Uruguay Round Agreement on Agriculture, held under the international General Agreement on Tariffs and Trade, now policed by the World Trade Organisation, has effectively limited and also signalled the eventual end of export subsidies and domestic market price protection. In addition, for the EU, its prospective enlargement to include the Central European countries, together with the unification of Germany, means that the traditional protection designed for an importing market is no longer viable. The EU support policy is now changing from one of price support to one of direct payments to producers, in return for more environmentally friendly land use practices and more animal friendly production systems. The consequence is that European red meat prices will become closer to their world market counterparts in time, and the competition between European supplies and those from elsewhere will tend to become more intense. The outcome of this competition, on top of the trends in domestic markets and marketing structures outlined above, will be considerable, if uncertain.

The UK red meat supply chain

Technical complexity of the supply chain

The red meat supply chain is among the most complex of all such chains. The primary producers are small independent businesses, frequently and perhaps increasingly pursuing their business in conjunction with a number of other activities and personal/family objectives. These producers utilise a number of different animal types and breeds, with a variety of feeding practices, feed sources and management systems, subject to considerable variation because of uncontrollable factors such as weather, disease and contamination. Furthermore, the finished fatstock is frequently not produced by the same business that is responsible for breeding and young animal production. There is a considerable trade in young stock (store animals) between farms and businesses, from those best suited for breeding to those best suited for finishing. This trade traditionally happens through livestock auction markets, which also deal in breeding stock, though is beginning to happen more frequently through direct contracts between regular buyers and sellers, and even via electronic auctions, though the latter depend critically on establishment of reliable quality and grading systems which prove especially difficult for breeding and store animals.

Finished animals arrive in the market in a variety of types, grades and conformations. For instance, just ten different animal types (breeds and cross-bred animals) fed through four different feeding regimes, producing three different carcass grades (because of uncontrollable variations in both nature and nurture) give rise to 120 different carcass types in the slaughterhouse. Furthermore, the final quality of these animals is highly time-dependent. Relatively marginal changes in selling times substantially affect both the costs of provision and the quality of the intermediate product, the animal carcass. Even the live animal assembly and transport systems can have significant effects on the quality and conformation of the final carcass.

Particularly for beef, the treatment of the carcass post slaughter is especially important for the final quality of the meat produced. Best quality beef needs to be hung and matured after slaughter as a carcass, for the necessary biochemical and biophysical changes to occur to produce tender and tasty meat. The conditions under which the carcass is hung and matured are also important. Differences in both hanging times and conditions further multiply the qualities and grades of the final carcasses.

The next step in the supply chain is the cutting and breakout of the carcass into joints and cuts, and the all-important by-products, offcuts and offals. Different carcass conformations yield substantially different proportions of the more valuable prime cuts and joints, and the secondary cuts, offcuts and offals. The value of the animal and its carcass depends critically on these proportions as well as on the quality of each. Once cut, the meat is then converted into final consumer products such as fresh, chilled or frozen boned or boneless joints, and cuts; pre-prepared fresh, chilled or frozen meat products (sausages, pies, hamburgers, etc.); and processed meats, including substantial use of offals and fillers (also appearing on the store shelves and in catering outlets as meat products, pet foods, fertilisers and hides such as leather and wool). The value added in this stage of the supply chain is critical to all involved. If this stage is mismanaged, then the commercial viability of producers, auction markets and traders, slaughterhouses, processors and packers is all potentially compromised, with the consequence that the whole supply chain becomes vulnerable. Not only that, but the health and safety factors in this complex can also break down for the smallest reason, and compromise the whole chain.

Finally, these widely different products are distributed and retailed through a variety of different outlets such as traditional butchers, supermarkets, corner stores and independent grocers, catering establishments of all sorts and sizes from pubs and fast food outlets and institutional (hospitals, prisons, offices, factories, schools and colleges, etc.) canteens to Michelin star restaurants. In short, to consider the red meat supply chain we need to examine a complex web of thousands of different individual supply and marketing networks. Its physical manifestation in plants and commercial operations is a reflection of this complexity.

Organisation and structure of the supply chain

Livestock auction marts, of which some 300 are operational in the UK, still account for over 50% of finished cattle sales and 70% of sheep sold liveweight. The balance is marketed through direct contracts, often on a deadweight basis, or through electronic auctions. The UK abattoir sector has experienced falling numbers of plants and rising throughput, both in total and *a fortiori* per plant. Total annual

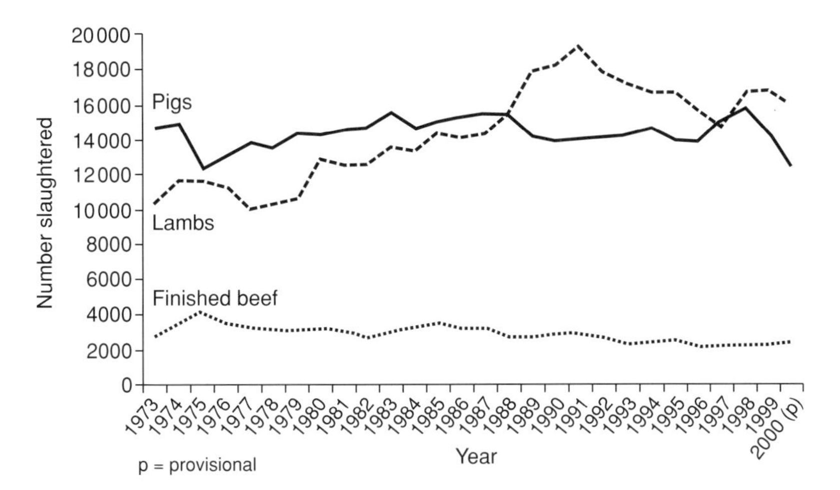

Figure 5.6 UK annual slaughter of finished animals (Source: DEFRA, 2002d).

throughput amounts to about 13 million 'cattle units (cu)' (one cattle unit is equivalent to one cow, or seven sheep or three pigs). There are now around 450 operational abattoirs in the UK, of which about 250 were full throughput plants in 1997, accounting for 95% of total throughput. Five per cent of UK throughput was accounted for by about 200 more small plants, with permanent derogations under EU health and safety legislation (of less than 1000 cu per annum). Seventy-two large plants, with a throughput of more than 50 000 cu, dealt with 76% of total throughput in 1996–1997. Many of the small abattoirs (less than 1000 cu) are connected with retail butchers' operations. Growth in direct transactions (slaughterhouse to retailer) has virtually eliminated traditional wholesalers from the abattoir–supermarket chain, though there is still clearly a need for wholesale distribution channels linking slaughterers, meat product manufacturers, smaller retailers and, more importantly, caterers. The history of UK slaughter of finished animals (excluding cull animals) is shown in Figure 5.6.

These annual slaughter rates are not frequently appreciated by the bulk of the population. The massive FMD outbreak of 2001–2002 resulted in the slaughter of 'only' 6 million animals in total (mostly sheep) versus a normal annual kill of about 32 million animals. Of course, the high visibility of open-air and seemingly brutal, if not unnecessary, slaughter and subsequent disposal obviously impacts quite differently on sensibilities than the normal slaughter behind closed doors and away from media attention. The full impacts of the exposure of this slaughter on consumption patterns and attitudes to the red meat sector are almost certainly still working through the chain, and cannot be expected to be favourable.

As can be seen in Figure 5.6, the most remarkable pattern has been the substantial rise in lamb production, despite the general fall in demand. This reflects the generally favourable rates of support for sheep production, especially in the less favoured areas (LFAs), resulting in a considerable increase in the UK sheep population, and also giving rise to growing concerns regarding overgrazing of some hill areas from an environmental point of view. This increase was encouraged during the early

1990s by the favourable sterling exchange rate, which has since strengthened to the disadvantage of the export trade, especially to France and the Middle East. This factor, together with falling prices and margins, exacerbated by the fallout from the BSE crisis and FMD, has resulted in the recent decline in domestic production.

Notwithstanding this general increase in throughput, overcapacity in the abattoir sector was still problematic (at perhaps 40% of peak kill), though it is difficult to separate out the critical capacity between the capacities of the unloading/dispatch bays, killing lines, boning halls, or chillers. Other things being equal, an increase in the proportion of the total market supplied from domestic sources would be expected to result in increased pressure on domestic slaughterhouse capacity. However, other things have not been equal. The restructuring of the supply chain complex towards a more concentrated flow of material to the major multiples, through centralised packing, processing and distribution operations, and away from localised and generally small-scale networks, has resulted in significant overcapacity amongst the smaller plants. This has been reinforced by new capacity investment to service the multiples' chains and to satisfy stricter health and safety requirements. More recently, the decline in domestic throughput has exacerbated the overcapacity situation, alleviated for at least some plants only by the enforced slaughter associated with the control procedures for BSE and more lately with the FMD outbreak.

Trade and the UK red meat supply chain

The market trends already outlined above are reflected in the development of the UK's self-sufficiency in red meats (Figure 5.7) that is generally increasing for both beef and lamb, especially the latter, until the downturn during the last few years as market prices fell and margins tightened, which was exacerbated by FMD resulting in suspension of exports during the crisis. The UK's sheep export trade has also relied to a considerable extent on the movement of live fat lambs into the French

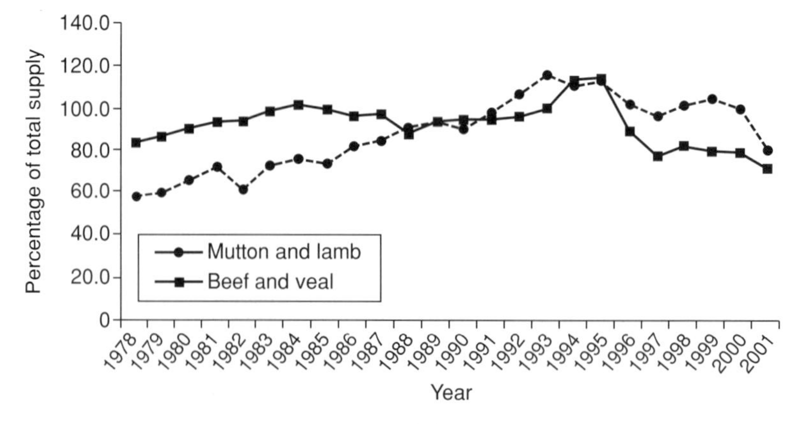

Figure 5.7 Domestic production as a percentage of total supplies (red meats) (Source: DEFRA, 2002c).

market, where demand is strong for 'French' lamb, and also for ethnic slaughter procedures. This live movement has attracted considerable adverse comment and protest, based on animal welfare concerns. The severity of the FMD outbreak, held to have been more difficult to control because of large-scale live animal movements, can be expected to lead to more restriction of such movements in the future; this will further weaken the UK's export markets and increase the over-supply situation on the domestic market.

Margins, price transmission and evidence of uncompetitive behaviour

The combination of the BSE crisis and the sustained pressure on domestic suppliers of red meats during the last five years of the past century prompted renewed allegations and accusations of uncompetitive or monopoly behaviour by the major retailers in the red meat chain. These accusations led to a number of independent studies of these behaviours. According to DEFRA (2001b), meat price changes (both increases and decreases) are fully or largely transmitted through the supply chain from producer to retail. However, there are lags in the price linkages, and these adjustments typically take several months to work through the chain. There is also evidence to suggest that retailers try to reduce the variability of retail prices which would otherwise arise from variability in producer and wholesale prices (so called 'levelling' behaviour). This study found no evidence to support the suggestion of a change in pricing behaviour within the supply chain during August or September 2000, when there were particularly strong allegations that retailers were failing to pass on reductions in producer prices. However, the analysis only considers changes in prices; it did not address the question of whether the absolute size of the margin between prices and costs is set at a competitive level.

The Competition Commission (2000) did not find that the retail multiples were generally behaving in an uncompetitive manner, despite their concentration and apparent domination of the market. In particular, referring to the DEFRA study, together with two other independent studies of meat pricing behaviour, they concluded (Competition Commission, 2000, p. 258) that,

> 'overall there was little dispute that in the wake of the BSE crisis, costs in the supply chain rose significantly and the overall value of the carcass fell because of the reduction in the value of some of the by-products. At the same time, consumer demand for beef and lamb fell, making it impossible for retailers to raise prices.'

The Commission also noted that the bulk of the effect of higher costs and lower value caused by the BSE crisis was felt by farmers, but considered (Competition Commission, 2000, p. 259) that, 'this is the outcome that would be expected in a broadly competitive market'. Unless retailers had previously been making excessive profits on fresh meat, for which there is no substantial evidence, they would only have been able to pay producers more by reducing their own margins to less than a normal rate of return.

Nevertheless, the Commission did find more general evidence that the major multiples were exploiting their undoubted buying power on occasions to the detriment of suppliers, especially smaller ones, and in particular those dependent on the majors for their returns. As a consequence, the Competition Commission (2000) provided a few suggestions covering primarily financial and essential business practices in that relationship and aiming towards the adoption of a Code of Practice between these chain members.

Overall, the Competition Commission felt there was considerable substance in many of the complaints by suppliers of bad or malpractice by the major supermarkets in their relationships with their suppliers. It is to be expected that there will continue to be substantial pressure on these powerful buyers (often of last resort) to behave more responsibly towards their suppliers. In the longer run, only those supermarkets that respect the rights and requirements of their suppliers can expect to maintain the integrity and cohesion of their all important supply chains. However, so long as suppliers are content to let these major buyers get away with bad practices, it is to be expected that the major multiples will continue to use them. One obvious remedy is for the smaller suppliers to co-operate and form selling consortia for their relationships with the major multiples. It is tempting to suppose that it is government's responsibility to ensure adherence to such Codes of Practice, and it may well be the case that continued evidence of bad practice amongst supermarkets could lead to a legally enforceable code being established in the future: perhaps this threat, and consequent additional costs of the associated bureaucracy and regulation of the marketplace, will be sufficient to encourage all links in the chain to improve their performance and behaviours without the full force of legislation. Indeed, this appears to be the case. The Department of Trade and Industry (DTI), in consultation with both supermarkets and suppliers, proposed such a Code of Practice in October 2001, and this was accepted by the four major supermarkets (Asda, Tesco, Sainsbury and Safeway) in December 2001, to become effective in March 2002 (DTI, 2001a,b).

Conclusions

The UK red meat supply chain complex at the beginning of the 21st century is clearly ripe for considerable change. The domestic market is oversupplied, and appears likely to remain so until and unless domestic production is reduced. The slaughter link, in particular, is plagued with over-capacity. In such conditions, margins at all stages in the chain will remain under severe pressure and considerable further 'rationalisation' can be expected. Several possible futures can be imagined for each of the links in the supply chain, though the critical links are at each end of the chain, the farms as the primary suppliers and the retailers as the final linkage to the consumers. The assembling, processing and packaging links can be expected to conform to the major driving pressures from each end of the chain. This is in contrast to most supply chains, where developments in primary production are largely driven by pressures from the consumer or final user end of the chain.

The reason why the primary production end of the red meat supply chain is important is that there are several other pressures influencing developments at the farm and land use end of the chain which are partly independent of the meat con-sumption drivers, exercised through the retailers. In particular, land will continue to be farmed in much of the UK, whether or not the particular products produced on this land can find profitable and commercial outlets. This is because at least some of the production systems (especially, perhaps, for red meats) are integral with the management of the natural rural environment, which is increasingly valued for its own sake. The ways in which land is farmed will increasingly depend on the landscape, wildlife and ecological characteristics that flow from land management practices. Farm production, at least in some areas and for some farms, is likely to become secondary. Nevertheless, there will be opportunities to market these products, though these will typically require differentiation of supply chains, rather than their further concentration and specialisation under the power of the supermarkets. This possibility poses particular problems for the analysis of future supply chain development.

Study questions

1. You are asked to provide an outline projection of the future condition of the UK red meat supply chain in 10–15 years' time. What key elements do you consider should inform this projection?
2. You are employed in the strategy group of a major multiple retailer in the UK. You are asked to prepare an options paper for meat and meat product retailing strategies over the next ten years. What options would you outline and why?
3. You are contracted to provide a beef producer group with options for developing their marketing chains. What options would you propose and why?
4. You have inherited a group of slaughter houses and meat processing plants in the UK, which currently comprise one large centrally situated slaughter house and packing plant specialising in white meats, and four smaller slaughter houses without processing and packing facilities currently operating below capacity and situated closer to the beef and sheep producing regions of the country. What are you going to do with this inheritance and why?

References

Altenborough, M. (2003) Consumer influences on MLC research strategy (http://www.iger.bbsrc.ac.uk/igerweb/stableton).

Colman, D. (ed.) (2002) *Phasing out Milk Quotas in the EU. Report to DEFRA, SEERAD, NAWAD and DARDNI.* Centre for Agriculture, Food and Resource Economics, School of Economic Studies, University of Manchester (http://www.defra.gov.uk/esg/reports/millequota).

Competition Commission (UK) (2000) *Supermarkets: a Report on the Supply of Groceries from Multiple Stores in the UK*. Competition Commission Report Series, London (http://www.competition/commission.org.uk).

Dairy Council (2001) *Dairy Facts and Figures*. Dairy Council, London.

DEFRA (June 2000) *Census*. The Stationery Office, London (http://www.defra.gov.uk/esg/).

DEFRA (2001a) *Milk Task Force Report*. The Stationery Office, London (http://www.defra.gov.uk/farm/mtfreport/mtfreport.pdf).

DEFRA (2001b) *Report of an Investigation into the Relationships between Producer, Wholesale and Retail Prices of Beef, Lamb and Pork*. The Stationery Office, London (http://www.defra.gov.uk/foodrin/compete/prices.pdf).

DEFRA (2002a) *Household Consumption of Selected Food Items*. The Stationery Office, London (http://www.defra.gov.uk/esg/).

DEFRA (2002b) *Farm Incomes in the UK*. The Stationery Office, London.

DEFRA (2002c) *UK Food and Farming*. The Stationery Office, London (http://www.defra.gov.uk/esg/m_natstats.htm).

DEFRA (2002d) *Monthly UK Slaughter Statistics*. The Stationery Office, London.

DTI (2001a) Press notice P/2001/606, 31 October 2001, Department of Trade and Industry, London.

DTI (2001b) Press notice P/2001/725, 18 December 2001, Department of Trade and Industry, London.

Food Act (1984) The Stationery Office, London.

Food Safety Act (1990) The Stationery Office, London.

Gallup (1995) *Responsible Retailing Initiatives: Report for the Co-operative Organisation*. Gallup, London.

Gunthorpe, M., Ingham, M. & Palmer, M. (1995) The meat industry. In *The UK Food and Drink Industry* (Strak, J. & Morgan, W., eds), Euro PA & Associates, Northborough, Cambs.

Harvey, D.R. (1999) Northern region agriculture and rural development: what futures? *Northern Economic Review*, **28**, 84–106.

Chapter 6
UK Crop Production

S. Wilcockson

Objectives

(1) To know the main indigenous crops involved, and the links that constitute the UK supply chain for different crop products.
(2) To realise the influence of factors such as agricultural policy measures, customers' quality requirements and crop assurance schemes on the supply chains and their management.
(3) To understand the structure of supply chains of different length and complexity and how they are currently managed.
(4) To appreciate that management may be improved by developing partnerships between different links in the chain from primary producer to final consumer to better meet their needs.

Introduction

This chapter aims to give a general picture of the characteristics and management of the diverse crop product supply chains in the UK. Because of the wide range of crops produced in the UK, in the main, only two of them, potatoes and sugar beet, are used to achieve the last two objectives. They are excellent examples of different types of crop supply chains for food for human consumption and they are very well documented. As everyone is familiar with some, if not all, of their products, the reader should be familiar with the end of the chain at least. Supply chains for other vegetable and salad crops are similar in many respects to potatoes. Consumers can buy them fresh, when they are collectively termed 'produce' by retailers. Some are processed to a greater or lesser extent, but processing is very highly developed in potatoes and a number of separate supply chains have evolved. In addition, the production of most of these crops is not subsidised or restricted by policy regimes, which has implications for supply chain management. Sugar beet, on the other hand, is completely different. The basic supply chain is even shorter and simpler, especially from the grower to the processor. Consumers invariably buy a fully processed product, such as granular sugar or syrup, or as a food ingredient in confectionery or soft drinks. Supply chains for other arable crops that have to be processed to a considerable degree before consumption are similar in many respects to sugar beet. Examples include, wheat for flour, bread and biscuits, barley for malt for whisky distilling and brewing beer, oilseed rape for cooking oil. In these cases, the grower is involved in fewer links in the supply chain and is more remote from the final consumer. Moreover, production of these crops is subsidised and controlled by means of the EU Common Agricultural Policy (CAP).

Crop production in the UK

The UK has a temperate, maritime climate. This is characterised by relatively mild winters, moderately warm summers, and rainfall that is evenly distributed throughout the year. Temperatures and rainfall vary from west to east and north to south, and are affected by local topography. Soils in which crops are grown include sands, clays, loams, silts and organic peats. These natural resources are the foundation for primary production that starts the supply chain from 'field to fork' and 'plough to plate'. They determine which crops can be grown, where and when, as well as their yield and quality. Most crops are grown outside in the open, but some are protected by modifying or controlling the natural environment. For example, tomatoes, cucumbers and lettuce are grown in glasshouses, often in soil-less media in hydroponics systems. Some field crops such as carrots and potatoes are covered with plastic and 'fleece' sheeting to increase temperatures, accelerate growth and advance harvest.

Major production statistics of cereal, oilseed, pulse and vegetable crops are shown in Table 6.1 and for potatoes and sugar beet in Tables 6.2 and 6.3. These are the arable crops that are grown mainly for human consumption and livestock feed, either in fresh or processed forms. However, with the trend to diversification, increasing amounts are grown for industrial uses for fuel, pharmaceuticals or fibres.

Growers sell or market most arable crop products for a cash return. Fodder crops are grown for feeding ruminant livestock, primarily cattle and sheep, and

Table 6.1 Combinable and vegetable crops in the UK (2000).

	Area (kha)	Yield (t/ha)	Total production (kt)	Average price (£/t)[a]	Total value (£million)[b]	Self sufficiency (%)
Wheat	2086	8.0	16 700	70	1580	117
Barley	1128	5.8	6490	70	685	135
Oilseed rape	332	2.9	965	134	246	86
Peas	67	3.7	247	79	39	n/a
Beans	124	3.9	485	80	74	n/a
Vegetables	137	—	2879	—	877	71

[a] Farm gate price.
[b] Including subsidies.
Sources: Department for Environment, Food and Rural Affairs (DEFRA), Scottish Executive Environment and Rural Affairs Department (SERAD), Department of Agriculture and Rural Development (Northern Ireland) (DARDNI), National Assembly for Wales Agriculture Department (NAWAD) (2002); Nix (2002).

Table 6.2 Potatoes in the UK (1995–2001[a]).

	1995	1996	1997	1998	1999	2000	2001
Area (kha)	172	178	166	164	178	166	166
Yield (t/ha)	37.4	40.7	43.0	39.1	40.0	40.1	39.4
Total production (kt)	6404	7228	7128	6422	7131	6652	6528
Average price (£/t)[b]	188	100	66	122	119	83	105
Total value (£million)	1095	636	390	630	750	454	600

[a] Figures for 2001 are provisional.
[b] Farm gate price.
Source: Department for Environment, Food and Rural Affairs (DEFRA), Scottish Executive Environment and Rural Affairs Department (SERAD), Department of Agriculture and Rural Development (Northern Ireland) (DARDNI), National Assembly for Wales Agriculture Department (NAWAD) (2002).

Table 6.3 Sugar beet in the UK (1995–2001[a]).

	1995	1996	1997	1998	1999	2000	2001
Area (kha)	196	199	196	189	183	173	177
Root yield[b] (t/ha)	43.0	52.4	56.6	52.9	58.0	52.5	46.1
Sugar yield (t/ha)			9.1	8.5	9.3	9.0	7.9
Total root production[b] (kt)	8 431	10 420	11 084	10 002	10 584	9 079	8 180
Sugar (% of roots)	16.5	18.0	17.2	17.3	17.2	17.1	17.2
Total sugar production (kt)	1 220	1 477	1 592	1 439	1 540	1 325	1 200
A/B quota beet price (£/t)	42	34	30	30	26.0	28.0	31.0
C quota beet price (£/t)	n/a	11.50	9.28	6.26	2.69	5.11	9.13
Total value (£million)	355	358	329	298	277	252	255

[a] Figures for 2001 are provisional.
[b] Adjusted to 16% sugar content.
Source: Department for Environment, Food and Rural Affairs (DEFRA), Scottish Executive Environment and Rural Affairs Department (SERAD), Department of Agriculture and Rural Development (Northern Ireland) (DARDNI), National Assembly for Wales Agriculture Department (NAWAD) (2002).

only a relatively small proportion is sold. Most is either fed *in situ* on the farm or conserved as silage or hay for use during winter, when fresh grass is not available. 'Set-aside' was introduced in 1992 under the CAP as a mechanism for managing the supply of certain crops. It is a percentage of the area of land capable of producing food crops that must not be used to grow a food crop if a grower is to be eligible to receive support payments from the EU. The largest area of arable crops produces an economic yield in the form of seed or grain. This includes the cereals, wheat, barley, oats and rye, oilseeds such as oilseed rape and linseed, and the pulses (peas and beans) that are harvested at maturity. They are also known as combinable crops because they are harvested with a combine harvester that cuts the crop and separates the seeds from the straw. Crops such as potatoes, sugar beet and vegetables are grown for their vegetative structures. Some of these are formed below ground, such as potatoes, sugar beet, carrots and other roots, or above ground, such as dry-bulb onions and vining peas for freezing or canning as 'garden' peas. These are often called non-combinable crops and require specialised, expensive equipment to harvest them. Most vegetable crops are sold fresh, but some are frozen, canned or dehydrated. Cereals, pulses and oilseeds for human consumption usually undergo some form of processing into a range of food products and ingredients.

The UK is about 80% self-sufficient in all indigenous, temperate crops, and imports the balance from the EU and other countries. Some commodities are exported and this is an important outlet in the supply chain of most crops. Supply chains for the major arable crops are illustrated in *Working Together for the Food Chain – Views from the Food Chain Group* (MAFF, 1999).

Common Agricultural Policy (CAP)

For many of the crops mentioned, the CAP manages the supply chain at the level of primary production. A system of import levies, export subsidies, intervention buying, quotas and quality specifications exerts some control over production, and hence prices received by UK producers, and supplies available for the supply chain. The UK joined the then European Economic Community (EEC), now called the European Union (EU), on 1 January 1973 and both agricultural output and self-sufficiency rose in response to higher prices, sometimes leading to overproduction. However, these controls are currently under pressure from the World Trade Organisation (WTO) negotiations for more liberalised world trade and access to EU markets for crops produced outside the EU. There is also pressure internally to decouple support from production to remove its distorting effects and improve competitiveness. Currently, the preferred option is to link support payments to environmental benefits (Clarke, 2002a,b).

Under the CAP, there is an Arable Area Payments Scheme (AAPS) which covers qualifying crops of cereals, oilseeds, protein crops and linseed. There is also the 'set-aside' requirement mentioned previously. It is a voluntary scheme designed to compensate arable growers for reductions in the support prices following reform of the CAP. Payments are linked to area of crops grown, not yield. Provided growers in England agree to set aside about 10% of the eligible area in 2003, they will receive payments ranging from about 371 euro/ha for cereals and oilseeds to 427 euro/ha for proteins. The more favourable rate for proteins is to encourage home-grown production, thereby reducing imports. At these rates, about 40% of growers' returns, gross margins or profits are subsidies in the form of Arable Area Payments.

For many of the non-combinable crops, however, there is no common EU regime and therefore no EU support. These crops include potatoes and most other vegetables. Sugar beet, cauliflowers and tomatoes are supported, but by mechanisms other than AAPS.

Details of the various support systems are too extensive and complex to consider further in this chapter. They are also subject to periodic modification and reform. Useful summaries can be found in Nix (2002) and Chadwick (2002) and full details in the Department for Environment, Food and Rural Affairs (DEFRA) and EU Commission documents.

Crop assurance schemes

The voluntary schemes for crops intend to assure the customer about the way in which the crop has been produced, stored and transported, and that certain basic standards have been met in accordance with the principle of due diligence. In particular, the schemes encourage integrated farm management (IFM). IFM aims to produce high quality, safe food at a profit for the producer that takes account of environmental and other responsibilities. For example, a whole range of approaches is used for crop protection and nutrition that are properly targeted and used carefully (LEAF, 2001, 2002). Producer members of the schemes are required to adhere to specific production protocols and record details of their agronomic management. They must undertake self-assessment and self-auditing procedures and be audited and validated by outside agencies. This approach gives transparency, traceability and accountability. The two main crop schemes are for non-combinable and combinable crops, respectively.

The Assured Produce Scheme (APS), established in 1997, was formerly the National Farmers' Union (NFU)/Retailer Partnership Scheme introduced in 1991. It covers about 40 different crops, including potatoes, fruit and vegetables, and is supported by the major supermarkets which require their suppliers to be members. The Assured Combinable Crops Scheme (ACCS) was also introduced in 1997 for grain and oilseed crops. Other schemes include Tesco's Nature's Choice which has additional requirements to the APS; Linking Environment and Farming (LEAF), an organisation promoting IFM, has recently launched the LEAF Marque, an environmental standard which can 'bolt-on' to existing assurance schemes, the logo of which will appear on food produced to the standard (LEAF, 2002). Some processors, such as the Unilever company Birds Eye Walls which freezes vining peas, require growers to be members of both LEAF and APS (Birds Eye, 2001).

Whilst assurance provides benefits for the customer, it does not give producers a premium. However, producers who are not assured may have difficulty in finding an outlet for their produce, and when they do they may have to accept a discounted price. For these reasons, although some of them feel that assurance schemes are unnecessary, bureaucratic and expensive, they seem to have little alternative but to join.

The British Farm Standard denoted by the 'Little Red Tractor' retail symbol on produce was launched in June 2000 and is administered by Assured Food Standards (AFS). It signifies that the food has been produced in accordance with assurance schemes and applies to both animal and crop products. There is general agreement that there is a need to rationalise the various schemes. The Food Standards Agency reported to the Policy Commission on Farming and Food in November 2001 following a review of concerns about low consumer recognition and understanding, confusion and the balance of costs and benefits. A key recommendation made to the UK government in January 2002 by the Commission was a possible rationalisation of all assurance schemes under the Little Red Tractor logo (Policy Commission on the Future of Farming and Food, 2002). The Food Standards Agency has produced a comprehensive review of food assurance schemes and, based on this, a position document for industry-wide consultation in September 2002 (Kirk-Wilson, 2002).

The potato supply chain

Potatoes, unlike sugar beet, cereal, oilseed and protein crops, are not currently subject to an EU regime. The potato industries, and in some cases the markets of individual member states, show considerable differences. In the UK, there are no price support mechanisms and the market is truly a free one, although this has not always been the case.

Control and management of the potato crop in the UK, at least at the primary production level, has changed markedly since 1997. Before then, and before the UK joined the EC, annual home production was controlled via the Potato Marketing Scheme that was administered by the Potato Marketing Board (PMB). Basically, the total area of production (and hence tonnage) was limited by quota (PMB, 1993): this control had been introduced to stabilise production. The yield of potatoes can vary considerably from year to year because of different weather conditions. For example, yields are lower in dry years (e.g. 1995), or when planting and harvesting is delayed by wet conditions (e.g. 2001). Variation in supply results in considerable fluctuations in price (Table 6.2). In the absence of production controls, high prices (low supply) would stimulate increased production in the subsequent year. If conditions and yields were more normal in this subsequent year, supply would increase and prices decline. The next year less would be grown and so on. Surprisingly, however, recent research has shown that this pattern is not evident in Belgium, although the structure of the industry is very different from that in the UK (Turner & Fearne, 2002). A major objective of the Potato Marketing Scheme was to avoid this year-to-year variability. Growers with more than 0.4 ha were required to register with the PMB, and the area that each could grow was limited. Each grower was allocated a basic area and the total of all growers' basic areas would be expected to provide 90–95% of home requirements, the remainder being supplied by imports. As yield levels increased due to improved varieties and crop management, or if the overall supply position required, a quota of less than 100% could be applied. For example, in a 90% quota year, a grower with a basic area of 10 ha would be eligible to grow 9 ha of potatoes (any area above 9 ha would attract an excess rate levy as a penalty). Since basic area was calculated as the average area grown over the last three years, a registered grower's basic area would decline over time. At national level, the decline in basic area would be balanced by the increase in national average yields and total supply would remain fairly constant.

The industry was deregulated, and in 1997 the Potato Marketing Scheme and PMB ended. However, there was support for a British potato organisation to continue some of the work of the PMB, and the British Potato Council (BPC) was established. Since 1997, there have been no area restrictions or intervention buying. Producers can grow as much as they like and the market is truly a free one.

Over the years, production has become very specialised and is becoming concentrated into the hands of fewer growers. The number of growers declined from 18 331 to 5606 or 70% between 1990 and 2000 whilst production showed a small change from 6.2 million tonnes (Mt) to 5.9 Mt (BPC, 2001). In parallel with the decline in producers has been a decline in the number of merchants, packers, processors and retailers.

The overall supply chain is relatively short and clear-cut. Growers produce potatoes for use as seed (about 10% of total production) or for the fresh market or for processing for subsequent purchase for consumption inside or outside the home. About 60% of all potatoes moving into human consumption are for the fresh market and 40% are processed. Frozen or chilled French fries account for almost 60% of all UK fresh potatoes that are processed, crisps about 40% and only a small amount (about 5%) is canned or dehydrated (BPC, 2002a). Producers may market their crops as individuals, especially if growing large quantities of potatoes, or as members of a co-operative or marketing group, and usually deal directly with buyers or via merchants or marketing agents. There are several supply chains including seed, pre-packing, processing, the fish and chip shop, and the more traditional 'loose potato' supply chain. A simplified diagram of the basic supply chain is shown in Figure 6.1. Primary producers supply one or more of the chains and the final consumer will purchase a range of different potato products from a range of different outlets. Obviously the chain is shortest where the producer and consumer are the same. This represents a small proportion of the national crop, including home-produced seed for planting the following year, or potatoes for own-consumption. The chain is also short where the grower sells potatoes direct to the final consumer at the farm gate, farm shop or farmers' market. However long the supply chain, customers at each link have their own specifications for the raw or processed potato and the aim is to add value at each stage. The final consumers buy potatoes and potato products for preparation in the home or for consumption in catering and food-service outlets such as restaurants and fast food counters. Most of the fresh potatoes are prepacked by packers such as Greenvale AP Ltd, MBM Produce Ltd, Solanum Ltd, the Higgins Group, Branston Potatoes Ltd, QV Foods Ltd, etc., for sale through supermarkets (Mossman, 2002). Currently about 80% of all fresh potatoes are washed and pre-packed into punnets or polythene bags of 2.5 to 5 kg weight. Most of these are sold through the major supermarkets such as Tesco, Sainsbury, Safeway, Asda and Morrisons. A relatively small amount is sold loose, including new potatoes and large 'bakers', and ever-decreasing amounts through wholesale markets and traditional greengrocers (Figure 6.2 a,b).

There is an expanding market for organic potatoes, and in 2000–2001 about 100 growers produced just over half a million tonnes on 2000 ha. Most organic

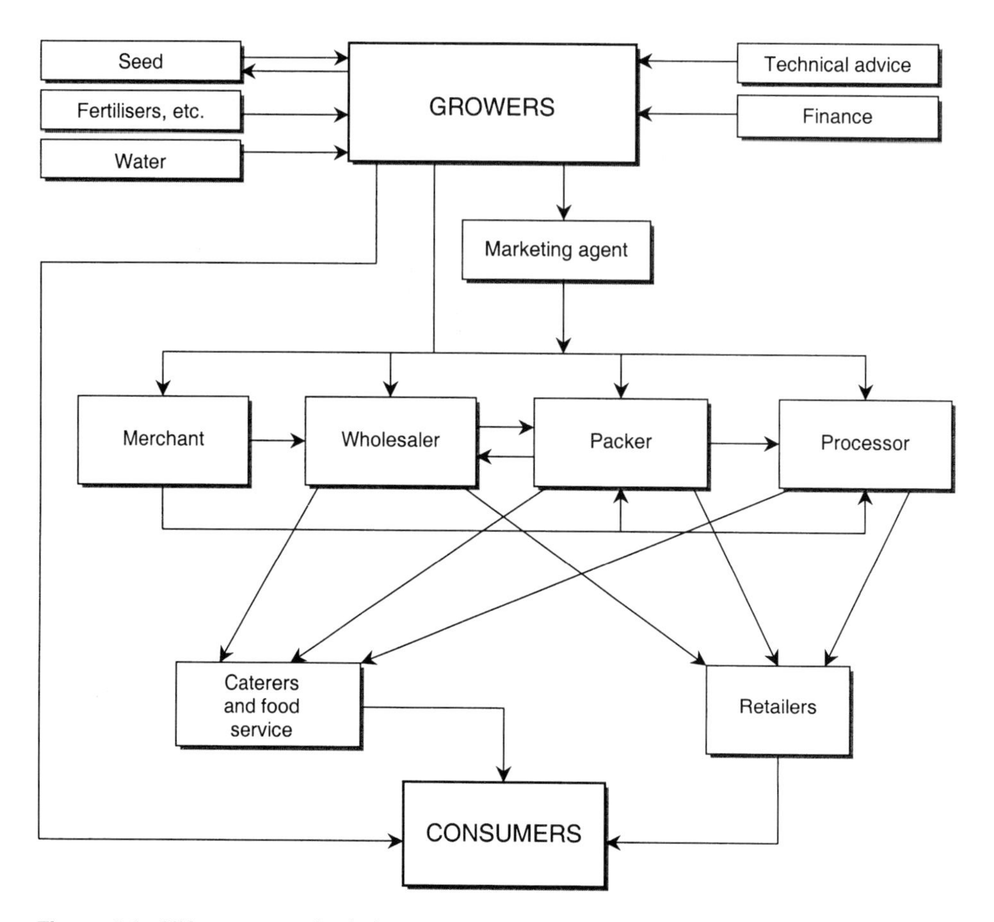

Figure 6.1 UK potato supply chain.

potatoes are sold fresh in pre-packs with only a small amount processed (BPC, 2002d). The market for processing potatoes is growing and, according to the British Potato Council's estimates, by 2005 more processed potatoes will be consumed than fresh (BPC, 2002e). Major French fry manufacturers include McCains and Potato and Allied Services, crispers Walkers (Fritolay), whilst KP Agriculture and Golden Wonder and Greenvale AP Ltd recently opened a dehydration plant. Dehydrated flakes and granules are used in instant potato products and snacks.

Some potatoes are grown on contract for processing because some of the requirements are different from those for the fresh market and so they have to be grown specially (NIAB, 2002). It may be difficult to obtain potatoes suitable for processing on a 'free-buy' basis on the spot market (i.e. not grown on contract). Potatoes for crisps and French fries should have a high dry matter content and low reducing sugar concentrations to produce crisps of the correct texture and colour. However, their cosmetic appearance is less important than it is for fresh market because they are peeled before processing and the consumer only sees the final, processed product.

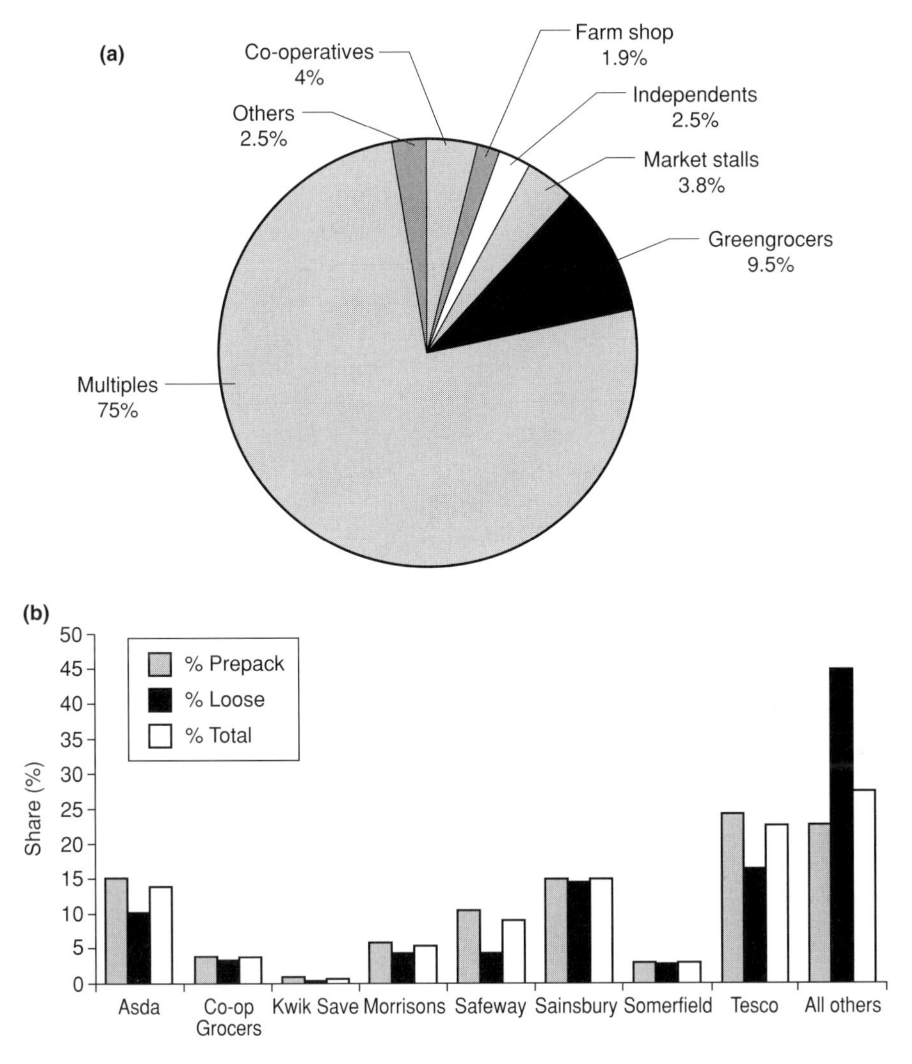

Figure 6.2 (a) Percentage market share of fresh potato sales outlets (2001–2002) (Source: British Potato Council, 2002b). (b) Retailer share of total fresh potato market by volume per cent (March 2001–March 2002) (Source: British Potato Council, 2002c).

On the other hand, pre-packed potatoes that are washed and packed in transparent bags need to have clear, unblemished skins. In the past, in contrast to processing, potatoes for the fresh market were not grown on contract, although there were some less formal agreements between growers and packers. Contracts for pre-packing have become more common recently, however. This is because potato production is becoming even more specialised with fewer growers and packers, all of whom make large investments to supply high quality material to very discerning and powerful retail outlets, and they take a considerable risk and expect definite commitments in return. In 2002, 57% of the crop area was intended for the spot market, but a large proportion of this was committed to a market through grower groups and regular buyers. The remaining 43% were grown on contract (20% for processing and 23% for ware and seed). For processing as a whole, 80% was grown on contract (31% of the ware and seed crop). Large growers planting more than 50 ha of potatoes intended to market over half of their crop on contract. For those with less than 10 ha, only about 15% was on contract (Gagen, 2002).

There are a number of different contractual arrangements between growers and packers or processors. These include fixed price contracts agreed at the beginning of the season, 60:40 contracts (60% of the crop at a fixed contract price and the remaining 40% at spot market price) or a market plus contract. In this case, the costs of producing potatoes are calculated and a price determined that provides a reasonable return. Supplies that do not meet the specific quality requirements may receive a price deduction or be rejected. Growers who store their crops receive additional payments the longer that they keep potatoes in store before unloading them for processing or packing, to cover the costs of storage and losses in store.

Traditionally, management of each link of this chain has been self-contained, i.e. growers have managed the growing crop, packers the packhouses, processors the processing operations, and retailers the distribution, promotion and sales to the final consumer. The marketing departments of packers and processors deal with suppliers, 'sourcing' raw material from both the UK and elsewhere. Packers also deal with wholesalers, direct with retailers, and also processors. This marketing expertise enables growers to market the whole of their crop, by supplying different qualities of their potatoes to different markets. It also offers their customers continuity of supply of potatoes of the right quality specification for processing or retail sale. Obviously, management at one of the links is influenced to a large extent by management of the link above and below and throughout the whole chain. At national level in the UK, overall management of the supply chain from primary producer to final consumer has been achieved by interaction between the individual links: the PMB and more recently the BPC have played a very important, overarching role. The BPC, financed by levies on growers and first purchasers of the crop, commissions research and development, provides technical information and promotes potato consumption. It also provides comprehensive and detailed UK and EU market information to facilitate decision-making and has dedicated potato supply chain managers (BPC, 2002f).

Recently there has been fusion of some of the links, shortening the supply chain even further as a result of mergers and take-overs, and many potato companies offer a fully integrated service (Mossman, 2002). For example, processors may supply the growers with high quality seed of the correct varieties, crop management protocols and technical advice, and assess the quality of the crop in the field and in the store. Processors may also have their own storage and transport facilities. For fresh potatoes, co-operation in the supply chain is increasing, particularly at primary producer level, with some formal co-operatives and a range of production and marketing groups with less formal arrangements. Packers work with some growers who supply on contract and others who are preferred suppliers but do not have a contract. They will also buy from producers who have no particular link with them at all, provided their raw material meets their specifications. Some packers grow their own potatoes. For example, Greenvale AP Ltd grows a large amount of organic potatoes for pre-packing through one of its dedicated lines. As well as packing and marketing crops, the packers also offer their growers various services such as seed supply and advice on agronomic management. These are also available to growers who do not supply that particular packer. The packers supply the supermarkets and at this level there is usually no specific, formal contract, but rather a 'preferred supplier' status. Supermarkets generally get their supplies from more than one fresh potato supplier and in the UK will buy both national and own-brand products from the processors.

The formation of partnerships, not necessarily formal agreements or contracts, between the various links in the supply chain is being further encouraged to benefit all involved from exchanges of information and expertise. The aim is to improve the match between supply and demand, to identify and consistently achieve the quality required by consumers in order to secure markets and gain from the 'added value' (Rickard, 2000; Warburton, 2002).

The sugar beet supply chain

Sugar in the form of sucrose is extracted from both sugar beet (a temperate crop) and sugar cane (a tropical crop). It is refined to produce white sugar crystals and a range of products. Most of this is used in the food and beverage industries, although it can be made into a range of chemicals, plastics, biodegradable detergents and biodiesel.

Sucrose was first obtained from sugar beet in 1747 by the German chemist Andreas Margraff. This potential source of sugar was not exploited until about 1811 by Napoleon, when cane sugar supplies to mainland Europe were cut off as a result of blockades during the Napoleonic wars, but had become dominant by 1880.

Sugar beet is a relatively new crop to the British Isles because requirements were supplied by imports of sugar extracted from sugar cane grown in the British colonies. However, for food security reasons, the industry began to evolve in the early 1800s, but slightly later than on mainland Europe. Factories opened in 1834 and 1870 but were short lived; one, built at Cantley in Norfolk in 1912, still operates today. Sugar cane continued to be an important source of sucrose, but because of supply problems during the early 1920s, the government subsidised production of sugar beet. By 1928, 18 factories had been built by independent sugar companies with the support and expertise of European sugar producers and refiners. In 1936 the industry was nationalised, the British Sugar Corporation took over, and on accession to the EC in 1973 became subject to the Sugar Regime. The British Sugar Corporation was privatised in 1980 to become British Sugar plc owned by Beresfords. The company was taken over by the Associated British Foods group in 1991. Since then, the industry has been streamlined and rationalised and the number of factories has declined from 18 to six (British Sugar, 2001, 2002).

Until joining the EEC, the UK obtained two-thirds of sugar requirements from sugar cane from Commonwealth countries under the then 20-year-old Commonwealth Sugar Agreement. This agreement lapsed on accession and was allowed to expire in 1974. The EEC then gave sugar quotas to the African, Caribbean and Pacific States (ACP) under the Lome Convention, which excluded sugar from Australia. The ACP quota was for 1.4 Mt of sugar. This amount was for imports into the whole EC, but as Europe had, and still has, a surplus of sugar from beet, it continued to come into the UK which was not self-sufficient in sugar. However, the UK sugar industry expanded from 1975 to 1980 to make up for the shortfall from Australia and is currently 60 to 70% self-sufficient.

British Sugar plc is the sole buyer of sugar beet in the British Isles and operates the processing factories located in eastern England and one in the west. The crop is subject to EU Regulation (EC) no. 2038/1993 governing community production of sugar that allocates quotas to different member states (Ekelmans, 1993). With one market for sugar beet, governed by quota, management of the supply chain is singular and easily identified. British Sugar plc has a fixed quota of sugar to extract from beet each year. It contracts with about 8500 growers to deliver about 8 to 10 Mt of clean beet containing about 16% sugar that is processed into about 1.2 to 1.5 Mt of refined white sugar (Table 6.3). Sugar beet is a popular crop with growers because it is bought at a fixed price and is profitable. British Sugar plc pays growers at the pre-agreed A/B quota price. This is based on the amount of sugar for which support is paid under the European Sugar Beet Regime and is higher than the world price. Sugar in excess of the A/B quota is termed C quota. This receives no support and is sold on the world market at the prevailing price and is a fraction of the value of A/B quota (Table 6.3). In 1999–2000, the total European supported tonnage of white sugar (i.e. A/B) was 13.76 Mt, of which the UK's allocated share was 1.144 Mt (i.e. 8.3%). The contract between the suppliers (the growers) and British Sugar plc is governed by the NFU/British Sugar Inter-Professional Agreement or IPA (British Sugar plc/National Farmers' Union, 2000). Growers have little opportunity to add value to the crop as it is bought at a fixed price, although there is, in effect, a bonus for high quality beet (based on sugar content of roots and impurity levels). Growers sell sugar beet to British Sugar plc for extraction and refining. Granulated white sugar is the main output and together with a range of products is marketed under the Silver Spoon brand-name.

The company supplies more than 50% of the total UK retail and food industry requirements and leads the industrial market, most of the balance being provided by Tate & Lyle plc, the sugar cane refiners. British Sugar plc's major industrial customers are sugar and chocolate confectioners, followed by soft drinks' manufacturers. Together they account for about half of its industrial sales. Bakers, biscuit makers and the canned and frozen food sectors account for most of the remainder (British Sugar, 2001). The sugar beet supply chain is shown in Figure 6.3.

Growers are contracted to British Sugar plc to supply a particular tonnage of beet, so that growers decide what area they need to grow on their farms to produce

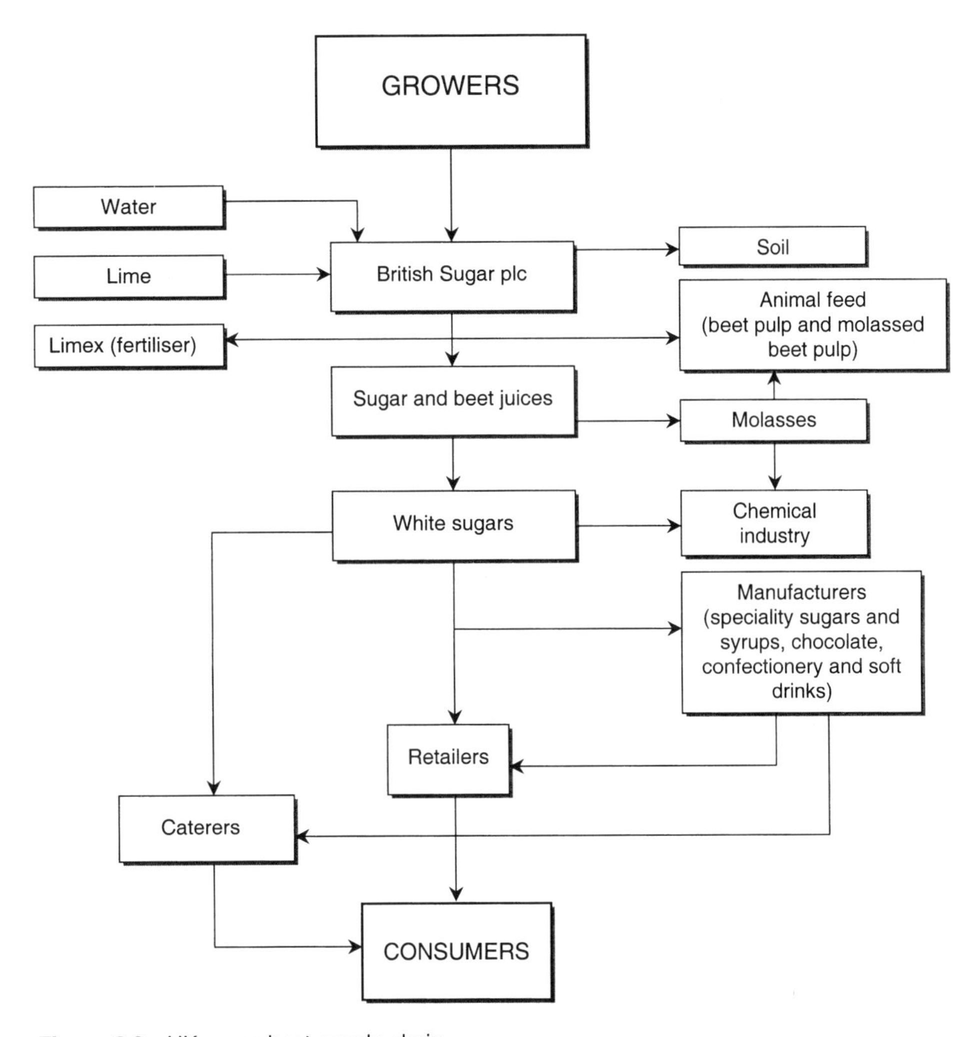

Figure 6.3 UK sugar beet supply chain.

this amount. Production of the crop is not covered by the Assured Produce Scheme (APS) because British Sugar plc operates its own quality assurance procedures.

As part of the contract, growers must buy their seed from British Sugar plc (one of Associated British Foods' companies is the seed processing company Germains). Throughout the growing season they are supported by area managers who give advice based on grower-sponsored research commissioned by the British Beet Research Organisation at institutions such as Brooms Barn and Morley Research Stations (King, 2002). The aim is for the individual grower and the industry as a whole to meet quota. If a grower fails to achieve contracted tonnage in two years out of three, his contract is reduced at first temporarily and then permanently if the shortfall continues. If the industry fails to meet its national quota for white sugar, then there is pressure from other EU member states for the UK quota to be reduced and reallocated to them. The factories are open from late September and process beet until January, and this is known as the 'campaign'. Growers supply their beet to factories according to basic quality requirements described in the contract. Individual growers deliver to a specific factory to ensure that, in total, the tonnage grown in each factory's catchment area is sufficient to fill the processing capacity throughout the beet-lifting season.

A permit system regulates the frequency of beet deliveries to achieve a steady flow of raw beet from the growers to the factories and avoid congestion on the surrounding roads. At the factory, the load of beet is sampled and soil and top tares, sugar and impurity contents are measured. Payments are then made on the basis of clean, correctly topped beet at 16% sugar content so that individual loads are 'adjusted' to 16% sugar. Effectively, beet with more than 16% sugar content gets a price bonus; beet below 16% sugar receives a price penalty. There is now a bonus for low levels of impurities (Bee, 2000). After extraction, the residue of the beet (pulp and molasses) is used to make animal feed. Lime, which is used in the sugar extraction process, and impurities from the beet juice are sold as a liming material for agricultural land under the Limex brand. Soil washed from the beet before processing is sold for landscaping. It is not returned to agricultural land because of the risk of spreading rhizomania. This is a serious disease of sugar beet caused by a virus that is present in a soil-borne fungus.

Not all beet are processed into white sugar immediately. To improve the efficiency of the factory operations some of the juice extracted from beet is stored in tanks as 'thick juice' for processing at a later date. There is currently interest in the production of organic sugar from home-grown beet and a demand of 1500 t by 2004 has been predicted. Because of increased costs of production and lower yields, growers are offered a 45% premium over conventionally produced sugar beet. In 2001, about 10 000 t of organic sugar beet was processed at the factory in Newark, UK, using a dedicated line (Jarvis *et al.*, 2001).

To improve management of the supply chain for both growers and processor, British Sugar plc introduced a 'one-off' outgoers' scheme in 2001. This allowed growers their first ever opportunity to permanently transfer sugar beet contract tonnage, independently of the sale of land, to other existing and new growers.

The aims were to increase competitiveness, consistently achieve full sugar quota, fulfil factories' processing capacity and reduce national haulage distances for beet from farms to the factories. It also allowed new growers to enter the supply chain and existing, more efficient growers to supply more beet by taking over contract tonnage from less efficient growers and others who wished to cease production (Bee, 2001). The scheme also supported British Sugar plc's 20:20 vision campaign to increase yields by 20% whilst lowering growing costs by 20% (Leeds, 2002). Such schemes and campaigns help to improve supply chain management and strengthen the position of the UK industry in the face of potentially far reaching changes in future. These include possible changes to the Sugar Beet Regime in 2006, the way the crop is supported (EC, 2000) and 'The Everything But Arms (Weapons) Treaty'. This treaty was first proposed in September 2000 as an EU Trade Development Assistance Scheme to allow duty free access into the EU for a range of goods originating from the world's 48 least developed countries. Under these proposals, sugar cane will come into the EU to be refined when the countries have an exportable surplus. The treaty will come into effect in 2007 at the end of the current EU Sugar Regime.

Conclusions

The theme of the Policy Commission on the Future of Farming and Food was reconnection; reconnecting farmers with their market and the rest of the food chain and reconnecting consumers with what they eat and where it has come from. A major recommendation in the report was to take measures to strengthen the food supply chain and promote collaboration among farmers. The UK government has responded by supporting the establishment of the Food Chain Centre to help build more effective and efficient food supply chains. This initiative should improve UK supply chain management, not only for crops, but also for all sectors of the industry.

Study questions

1. What are the main crops produced in the UK and what are they used for?
2. How does the CAP affect the food supply chain for crops and what changes are there likely to be in the future as it continues to be reformed?
3. What are the main assurance schemes for crops and what are their objectives?
4. Why is supply chain management more complicated for potatoes than for sugar beet, and how can supply chain management be improved?

References

Bee, P. (2000) The sugar beet contract – a new deal. *British Sugar Beet Review* **68**, 4–8.

Bee, P. (2001) Sugar beet outgoers' scheme – the results. *British Sugar Beet Review* **69**, 2–4.

Birds Eye (2001) *In Pursuit of the Sustainable Pea. Forum for the Future*. Birds Eye/Unilever, London.

BPC (2001) *Potato Statistics for Great Britain: Historical Data – 1960 Onwards*. British Potato Council, Oxford.

BPC (2002a) *Potato Consumption and Processing in Great Britain 2000/2001*. British Potato Council, Oxford.

BPC (2002b) GB retail potato summary. Distribution of fresh potato sales volume by outlet. In *Potato Statistics in Great Britain (1997–2001)*. British Potato Council, Oxford.

BPC (2002c) *Retailer Share of Total Fresh Potato Market Volume and Value*. Retail report, GB potato summary April 2002. British Potato Council, Oxford.

BPC (2002d) Organic potato production. British Potato Council, Oxford (http://www.potato.org.uk/).

BPC (2002e) *Understanding Consumers – Vital for the Future of GB Potatoes*. BPC Understanding Consumers' Research Programme – Factsheet 1. British Potato Council, Oxford.

BPC (2002f) *About the British Potato Council*. British Potato Council, Oxford (http://www.potato.org.uk/).

British Sugar (2001) *Information – Facts about British Sugar 2001/02*. British Sugar plc, Peterborough.

British Sugar (2002) *About British Sugar*. British Sugar plc, Peterborough. (http://www.britishsugar.co.uk/).

British Sugar plc/National Farmers' Union (2000) *Inter-professional agreement*. British Sugar plc, Peterborough.

Chadwick, L. (2002) *The Farm Management Handbook*, 22nd edn, September 2001. SAC, Edinburgh.

Clarke, P. (2002a) Mid-term review: one capped cheque is what awaits UK farms. *Farmers Weekly* **137**(2), 14–15.

Clarke, P. (2002b) Fischler's CAP plans under fire, *Farmers Weekly* **137**(3).

DEFRA, SERAD, DARDNI & NAWAD (2002) *Agriculture in the UK 2001*. The Stationery Office, London.

EC (2000) *Proposal for a Council Regulation on the Common Organisation of the Market in the Sugar Sector*. Commission of the European Communities, 4/10/2000. COM (2000) 604 final, 2000/0250. European Commission, Brussels.

Ekelmans, J. (1993) Sugar and isoglucose. In: *The New Regulation of the Agricultural Markets*, pp. 21–27. Vademecum, Green Europe.

Gagen, G. (2002) *Marketing Intentions 2002, Report for Great Britain*, British Potato Council, Oxford.

Jarvis, P., Leeds, S. & Cormack, W. (2001) Organic sugar beet production in the UK. *British Sugar Beet Review* **69**, 10–11.

King, J. (2002) The levy-funded research and technology transfer programme for 2002/03, British Beet Research Organisation (BBRO). *British Sugar Beet Review* **70**, 42–43.

Kirk-Wilson, R. (2002) *Review of Food Assurance Schemes for the Food Standards Agency*. Food Standards Agency, London.

LEAF (2001) *Linking Environment and Farming (LEAF): Annual Review 1999–2000*. National Agricultural Centre, Stoneleigh.

LEAF (2002) *About LEAF*. National Agricultural Centre, Stoneleigh (http://www.leafuk.org).

Leeds, S. (2002) 20:20 vision and CPI (crop profitability initiative) 2001. *British Sugar Beet Review* **70**, 14–19.

MAFF (1999) *Working Together for the Food Chain – Views from the Food Chain Group*. Ministry of Agriculture, Fisheries and Food, London.

Mossman, D. (2002) Buyers' need for volume and continuity will speed trend to fewer units. *Agronomist* **1**, 7–8.

NIAB (2002) *2002 Potato Variety Handbook*. National Institute of Agricultural Botany, Cambridge.

Nix, J. (2002) *Farm Management Pocketbook*, 32nd edn, Imperial College at Wye, Ashford.

PMB (1993) *Basic and Quota Area for Potatoes – A Guide for Producers and their Advisers*. Potato Marketing Board, Oxford.

Policy Commission on the Future of Farming and Food (2002) *Farming and food – a sustainable future*, Report of the Policy Commission on the Future of Farming and Food. Cabinet Office, London.

Rickard, S. (2000) Potato sector leads the way. *Challenges and prospects*, Vol. 4, LloydsTSB Business, London.

Turner, P. & Fearne, A. (2002) *Do We Measure Up? A Review of the Potato Processing Industries in Britain, Netherlands, Belgium and France*. British Potato Council, Oxford.

Warburton, R. (2002) *Boxing clever – The Challenges Facing Processors and Packers. Yield.* **1**(1). Bidwells Property Consultants.

Chapter 7
Food Manufacturing

D. Hughes

Objectives

(1) To describe and outline the importance of the food- and beverage-manufacturing sector in the overall operations of the food supply chain.
(2) To explore initiatives that large-scale food manufacturers can consider to respond to the growing power of retailers in the food supply chain.
(3) To identify initiatives that small- and medium-scale manufacturers can consider when establishing commercial strategy to compete in a food industry increasingly dominated by large-scale players.

Role of manufacturing in the UK food supply chain

The principal role of manufacturing in the UK food supply chain is to 'add value' to food product ingredients such that they meet the requirements of other supply chain members, including consumers themselves. The act of undertaking manufacturing activities involves adding costs and the *de facto* commercial rule is that the value added should, at a bare minimum, exceed the costs incurred. If not, why bother doing it?!

In history, manufacturing activities were undertaken by household members and specialised firms, largely with a view to preserving otherwise perishable food products between seasons for crops, and between slaughtering for meat products. Basic preservation techniques, such as drying, smoking, salting, pickling and, later, bottling, canning and freezing, are still today the bed-rock techniques of food manufacturing, although new techniques have emerged, such as (controversially) irradiation, modified atmosphere packaging (linked to high integrity chill chain transportation and storage), and sous-vide (cooking under vacuum) meal preparation. Envisage a manufacturing continuum with an early hominid smoking strips of meat on a fire at the mouth of the family cave at one end and a Northern Foods plc processing plant spinning out 'restaurant quality' Marks & Spencer's Café Culture chilled ready meals at the other.

One characteristic of the nature of food products in many, particularly economically developed, countries has been the burgeoning and inexorable growth in demand for convenience, 'ready-to-eat', 'ready-to-heat', etc., attributes. This does not reflect consumer indolence, but rather changing lifestyles where the principal meal preparation constraint is time, not money. Meal preparers are willing to pay others – decades ago servants, now food manufacturers or food service providers – to strip the labour and inconvenience out of feeding themselves and their families. In short, the raw ware potato has become the problem, and the microwaveable French fry the solution.

The conclusion is that the role of the food manufacturer will increase in importance. An important implication for farmers is that, increasingly, they are producers of low-priced raw materials to which others add significant value.

Currently, manufacturers do much more than simply transform ingredients into finished food products. They have a key role in linking the point of manufacture with the point of product sale (supermarket, corner store, etc.) and, in history, have been leading transportation providers. Transportation of final products to point of sale accounts for approximately 40% of total distribution costs, with warehousing responsible for much of the remainder. However, the manufacturer's responsibility for transportation of goods to the major supermarket chains that account for 85% or more of the grocery trade may be increasingly a function of the past.

The major chains see the logistics function as a significant area for future cost savings. For example, Sainsbury's Supermarkets Ltd have tasked Exel, a large logistics provider, with managing Sainsbury's primary logistics, i.e. the leg between supplier and the supermarket's regional distribution centres (RDCs) that are centres from which the Sainsbury store network is presently serviced. The next logical step will be to link primary and secondary distribution, from point of manufacture to point of retail sale. In effect, supermarket chains will take control of the supply chain at the 'factory gate'. One implication of this development is that retailers are and will increasingly be asking manufacturers to quote product prices on a factory basis. This is referred to as 'ex-factory gate pricing' and, as it emerges in 2002–2003, is viewed with trepidation by some manufacturers, not least those who have made significant investments in distribution plant and facilities.

The importance and structure of food manufacturing in the UK

The food manufacturing sector has a substantial ambit, as is reflected in its significant importance in the food chain *per se* and overall UK manufacturing. Take the commodity milk, for example: the 'living' manufacturing plant, the cow, transforms grass into raw milk. Dairy processors may pasteurise, homogenise, differentiate by fat content and package milk for sale on the 'fluid' market, produce butter, cheese, milk powder, yoghurts and other dairy desserts for sale to retailers, and/or service the requirements of food ingredient manufacturers for specialised milk proteins and milk by-products such as whey. Thus, the food-manufacturing sector is not simply one substantial link in the food chain: it is a series of linkages comprising primary processors, who have an important assembly function. They may split basic commodities into component parts for sale to more specialised food (and, indeed, non-food) manufacturers who, in turn, may deal with firms that repackage bulk products into consumer-size units or formats that meet the requirements of the food service trade.

In terms of value of production, the European Union (EU) food and drink industry accounts for 15% of all production in manufacturing in the 15 countries of the EU (EU-15). It has been reported that, 'With a total production of 593 billion euro in 2000 and an added value of 133 billion euro, the food and drink industry is ranked first ahead of the chemical, car, machinery and equipment industries. That sector is the third largest industrial employer in the EU (12% of the EU industrial workforce)' (USDA, 2002) with France and Germany in first and second place, respectively. The importance of the industry in the EU is reflected similarly in the UK manufacturing economy, although it is the relatively strong contribution of the UK drinks sector (alcoholic and non-alcoholic) that helps explain its buoyant EU performance (pushing Italy and Spain into fourth and fifth place, respectively).

UK food manufacturing sales for 1996 and 1999 by principal product categories are shown in Table 7.1. The grain-based sector (bread, biscuits, etc.), followed by meat, beer and soft drinks, and then dairy processing categories, top the sales list. The total sales for 1996 and 1999 (£47 bn and £46 bn, respectively) show that the industry overall has struggled in recent years to achieve any 'top line' growth; indeed, the industry has experienced deflationary pressures. This reflects factors such as: slow-to-static population growth and ageing population, the increasing buying power of fewer, larger supermarket chains who have embraced every-day low price (EDLP) merchandising strategies with the promise to their shoppers of 'always lower prices', and the globalisation of supply sourcing. To emphasise the deflationary struggle that has faced the food manufacturing industry in recent years, it is instructive to focus specifically on year-on-year sales of fast-moving manufacturer-branded, food consumer goods (FMCG) sold via retail in the UK (Information Resources, 2002):

- Year-on-year sales (by value) of manufacturer-branded food and non-alcoholic drink products from 24 retail categories increased by 2.4% between 2001 and 2002, i.e. sales barely kept up with inflation, in spite of significant investment in above- and below-the-line promotion by FMCG food firms. In 2001, the expenditure on advertising for the top twenty branded food and beverage products exceeded £125 million (see Table 7.2).
- Some food product categories did better than others: manufacturer-branded frozen food products saw no growth in sales value between 2001 and 2002, whereas the chilled meals category, dominated by retail brands, grew by 18% year-on-year. Reasons explaining such differential performance include consumers' perception of chilled foods as being of premium quality compared with frozen foods, ready meals being unequivocally in-tune with contemporary lifestyles, and strong demand for high levels of convenience (Information Resources, 2002; Taylor Nelson Sofrés, 2002).

Table 7.1 Food and beverage manufacturers' sales (£m) by principal product categories in the UK for 1996 and 1999 (Office of National Statistics, 2002).

	1996	1999
Meats		
Production and preserving red meat	3 650	3 314
Production and preserving poultry meat	1 675	1 601
Bacon and ham	850[a]	874
Other meat and poultry meat processing	3 452	3 691
Sub-total	9 627	9 480
Fish		
Processing and preserving fish products	1 531	1 532
Dairy operations and ice cream		
Dairy operations	5 794[b]	5 593
Ice cream	567	484
Grain products		
Breads, pastries, cakes	3 785	3 877
Biscuits, preserved pastries	3 249	3 062[c]
Grain mill products	2 927	2 707
Sub-total	9 961	10 130
Fruit, vegetables and potatoes		
Processing and preserving potatoes	1 148[b]	1 225[a]
Processing and preserving fruits and vegetables	2 005	1 981
Fruit and vegetable juice	488	473
Sub-total	3 641	3 679
Food products		
Cocoa, chocolate and sugar confectionary	3 139	3 323
Sugar	1 383	1 205[c]
Processing of tea and coffee	1 349	1 484[c]
Other food products not specified elsewhere	1 800	1 871
Sub-total	38 792	38 297
Drinks[d]		
Beer	5 458	4 719[c]
Mineral water, soft drinks	2 700[a]	2 718
Sub-total	8 158	7 437
TOTAL	46 950	45 734

[a] Estimate.
[b] 1997 figures.
[c] 1998 figures.
[d] Does not include spirits.

- Commodity categories, i.e. food product categories where shoppers perceive little differences between the range of products offered at retail (e.g. frozen French fries), have come under particularly strong price pressure as year-on-year sales values were static, although sales volume (units of product sold) increased significantly.
- Reflecting the changing structure of food manufacturing in the UK, in most major food product categories, two or three manufacturers dominated sales of branded goods, e.g. Walkers products (a subsidiary of US-registered PepsiCo)

Table 7.2 Top twenty British food and beverage brands in 2001[a] (ACNielsen, 2001).

Rank	Brand	Manufacturer	Sales (£m)	Advertising spend (£m)	Advertising (as percentage of sales[b])
1	Coca-Cola	Coca-Cola	491	19.5	4.0
2	Walkers Crisps	PepsiCo	383	10.5	2.7
3	Nescafé	Nestlé	312	8.2	2.6
4	Müller desserts	Müller	306	11.0	3.6
5	Kingsmill bread	Allied Bakeries	192	2.7	1.4
6	Robinsons	Britvic	187	6.8	3.6
7	Hovis bread	RHM	172	3.8	2.2
8	Whiskas	Mars	170	8.5	5.0
9	Warburtons bread	Warburtons	162	1.8	1.1
10	B. Matthews meats	B. Matthews	155	1.2	0.8
11	Pedigree	Mars	148	6.9	4.7
12	McCain Chips	McCain	148	5.0	3.4
13	Kit Kat	Nestlé	145	11.0	7.6
14	Birds Eye fish	Unilever	143	N/A	N/A
15	Young's fish	Young/Bluecrest	140	2.7	1.9
16	Flora	Unilever	133	9.0	6.8
17	Birds Eye poultry	Unilever	133	0.2	0.2
18	Heinz soup	Heinz	130	N/A	N/A
19	Heinz beans	Heinz	129	N/A	N/A
20	PG Tips	Unilever	127	5.6	4.4

[a] Year ends August 2001.
[b] Unweighted average 3.2%.
N/A: not available.

Table 7.3 Food and beverage manufacturers[a] in the UK by employment size, 2001 (Office of National Statistics, 2002).

Employment size (persons)	Number of manufacturing units
1–9	6 070
10–19	1 725
20–49	1 180
50–99	600
100–199	495
200–499	455
500–999	130
>1000	45
TOTAL	10 700

[a] Includes tobacco.

accounted for over 50% of bagged snack sales; Mars, Cadbury and Nestlé were paramount in chocolate confectionery, McCain in frozen potato products (Information Resources, 2002).

There are more than 10 000 food and drink manufacturing firms in the UK, of which 73% employ less than 20 persons, and 6% 200 persons or more (Table 7.3). In an era of growing supermarket chain concentration, coupled with a predilection for the chains to source products from fewer, larger suppliers, and to constrain the manufacturer-branded product offer to primary and secondary brands, the larger manufacturers account for a disproportionately high percentage of food and drink turnover at retail. The UK food manufacturing industry is more concentrated than the EU average: in 1990, UK large enterprises accounted for 66% of total output, compared with the EU average of 40%. Over the past few decades, there has been a continuing trend to more concentration, although change has not been uniform, with some sectors increasing and others decreasing (Gilpin & Traill, 1995). Even by the early part of the 1990s, however, the top three manufacturers accounted for more than 50% of output in the pasta, coffee, biscuit, chocolate confectionery and mineral water product sectors in most EU countries (Hughes & Ray, 1999).

Taking an international perspective, the incidence of mergers and acquisitions (M&A) within the global food industry increased during the 1990s and early 2000s. While the globalisation of retailing is a relatively recent phenomenon, food manufacturing has been international in character for decades. Globally, food manufacturing is led by firms with their headquarters in the USA. The instigator of the convenience food revolution, the USA, accounts for about one-quarter of the industrialised world's processed food production (Henderson *et al.*, 1996). The value of large-scale transactions, i.e. acquisition deals of £300 million or more, is slowing as prime targets are picked off. In Table 7.4, a synopsis of the more significant mergers and acquisitions completed during 2001 and 2002 is presented. The list is dominated by American firms and Nestlé, and illustrates that global firms are either selling off non-core parts of their business (e.g. Unilever divesting the bakery

Table 7.4 Major international food and beverage company acquisitions[a] (2001, 2002) (Rabobank International, 2002).

Target company	Bidder company	Product group	Acquisition value (US$ million)
Quaker Oats	PepsiCo	Snacks/soft drinks	14 392
Ralston Purina	Nestlé	Pet food	10 700
Keebler Foods	Kellogg	Snacks/cereals	4 652
Earthgrains	Sara Lee	Bakery	2 800
IBP	Tyson	Meat products	2 700
Chef America	Nestlé	Convenience snacks	2 600
Bestfoods Baking Co.	Unilever	Soup/sauces	1 765
Kamps	Barilla	Bakery	1 610
Schöller	Nestlé	Ice cream	1 185
EU soup and sauce business	Campbell	Soups/sauces	950
Galbani cheese and meat unit	Danone	Dairy/meat products	915
Van Melle	General Mills	Ice cream	865
Ice cream Partners USA	Nestlé	Ice cream	641
Minnesota Corn Processors	Archer Daniel Midlands	Cornstarch and oil	636

[a] Deals over US$600 million.

part of its recent acquisition Bestfoods), or purchasing companies which have strength in core categories for the purchaser (e.g. Nestlé buying Ralston Purina for its pet food business, and the ice cream element of General Mills).

USDA research shows that foreign direct investment (FDI) has become an increasingly important strategy for major food companies, not least those from the USA, to expand abroad. American firms, such as Philip Morris, ConAgra, PepsiCo, Coca-Cola, Cargill, Tyson, Mars, Heinz, Sara Lee and Kellogg, and European firms, such as Unilever, Nestlé, Danone, Parmalat and Cadbury Schweppes, have used FDI to invest capital in overseas production rather than ship the product from a domestic source. Bolling & Somwaru (2001) state that, 'Companies use FDI to circumvent trade barriers, gain access to less expensive resources, and tailor products to local tastes in other markets. These factors are especially important to the processed food industry'.

Even with much media talk about globalisation, food tastes, even within a modest geographical area such as the EU-15, do differ significantly, thus reducing the latitude for international food firms to build global brands. This provides opportunities for small- and medium-scale food manufacturing enterprises to exploit local/regional 'hero' brands, which may be increasingly available if larger firms seek to slim down their brand portfolios and concentrate on branded products which have pan-territorial potential.

Yet many of the principal food and beverage brands in the UK are owned by multinational companies that are globally dominant in the categories in which they are most focused (Table 7.2), for example, Mars in chocolate confectionery and pet food, Unilever in frozen food products and margarines/spreads, Coca-Cola in cola and other non-alcoholic drinks, PepsiCo in snack foods. These companies have built impregnable market positions over decades of investment in marketing and research and development (R&D). Indeed, only 10% of the top 100 grocery brands in the UK (based on sales) have been launched in the last ten years whilst 65% were introduced to the British market 30 years or more ago (Figure 7.1). An indication of the future direction of global alliances in the food industry is provided by Nestlé's

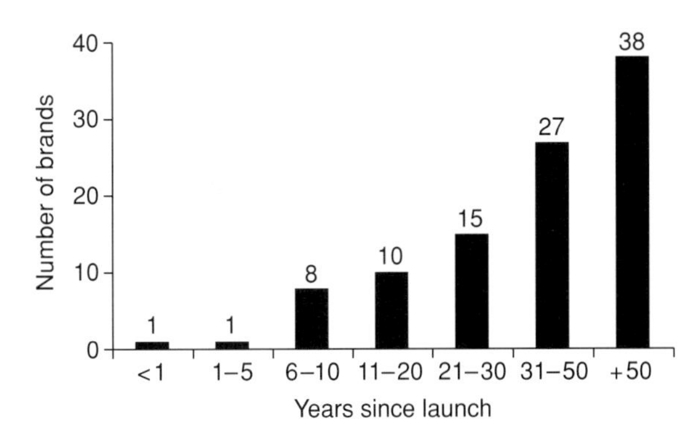

Figure 7.1 Top 100 grocery brands launched in the UK in the past 10 years (Source: Taylor Nelson Sofres, 2002).

joint venture link with the cosmetic company L'Oréal in July 2002. The two global corporations are co-operating to produce food products with cosmetic purposes (e.g. softer, wrinkle-free skin, etc.). We are what we eat!

Large-scale food manufacturers: responding to increasing concentration in grocery retailing

Even for the major fast moving consumer goods (FMCG) food companies, markets in the UK, EU and further afield provide a tough competitive environment, particularly the low growth, mature markets of the developed world. From a period of market dominance in the 1950s and early 1960s, manufacturers have come under increasing commercial pressure from major retail chains who are now expanding aggressively on a global basis. In their home markets, food companies have learnt that strong brands and market leadership are critical to fending off the challenge of retailer own-label products. The UK is a particular stronghold of sophisticated private/own label products; most chilled prepared food products are retailer-branded, and for packaged frozen and shelf stable grocery products, the three market leaders have private label shares exceeding 40% (Figure 7.2). Further, the major retailers are fast developing sub-brands: premium labels such as Tesco's Finest and Sainsbury's Taste the Difference are cases in point. Of course, retailer brands present both opportunities and threats to manufacturers. On the one hand, they represent an opportunity to produce on behalf of retailers and expand production volumes from existing processing plants; on the other hand, retailer own-label manufacturing is consistently lower net margin business than producing manufacturer branded products (Figure 7.3).

There are a range of responses that FMCG companies can consider as mechanisms to combat the growing competitive threat from increasingly powerful global retailers, such as:

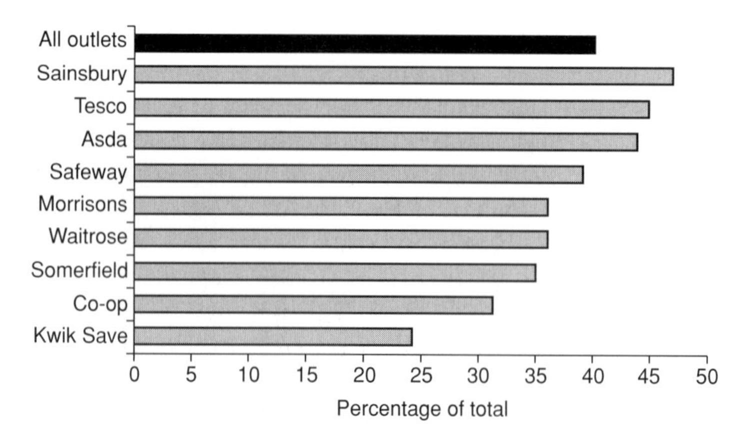

Figure 7.2 Packaged UK grocery private label levels in 2000 (Source: Taylor Nelson Sofres, 2002).

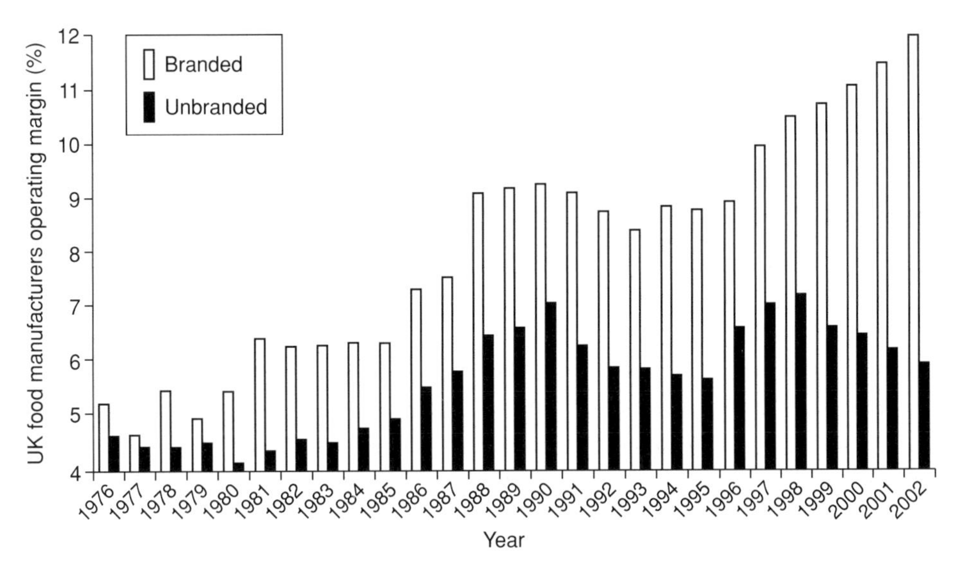

Figure 7.3 UK branded and unbranded net operating margins for food manufacturers, 1976–2002 (Source: Investec Henderson Crosthwaite, personal communication).

- *Innovate*: In general, retailers' own label products are copies of manufacturer brands, and retailers have no history of significant investment in new product development (NPD). FMCG firms have the resources to invest in a high level of R&D expenditure to develop products which have proprietary technology. At 1.5% of sales, R&D expenditure in the international food industry is at very modest levels relative to the 14–15% levels of the pharmaceutical and biotechnology industries (DTI, 2002), although within the food manufacturing category, some FMCG firms invest substantially more than average, for example in 2001, Unilever invested 2.3% of sales in R&D. It is, perhaps, salutary to note that those same pharmaceutical and biotechnology firms who are investing 15% of sales in R&D are those that are most active in the life science field. In time, these investments will result in the launch of proprietary-owned functional foods/nutriceuticals and will serve to alter the power balance within the food industry.
- *Cost leadership*: Through the 1990s, firms world-wide have sought to 'downsize' or 're-engineer' to cut costs and remain competitive. Notwithstanding the plausibility of Porter's generic business strategies (Porter, 1980), cost consciousness and leadership are necessary conditions for market leadership. However, another Porter concept, value chain analysis, is apposite (Porter, 1985). It is counterproductive to cut costs and, through so doing, reduce customer value by an even greater amount. The challenge is to identify what customers value and are willing to pay for, and then to focus on the key drivers of cost advantage, i.e. economies of scale and of learning, production techniques, product design, input costs, capacity utilisation, and managerial/organisational efficiency and their minimisation, whilst keeping customer requirements firmly in sight.

- *Geographic diversification*: One challenge for manufacturers over the next decade will be to establish the primacy of their brand names in new market territories before the retailers establish their own.
- *Identification of new consumer needs and usages*: This activity is an extension of innovation. An example from the 1990s includes the Campbell Soup Company re-launching its canned soup product range to take advantage of growth in the pour-on sauce sector. In the UK in the 2000s drinking yoghurt has been created as a new category and exploited by, *inter alia*, Yakult, Danone and Nestlé, promoting their products on the basis of healthy lifestyles and, specifically, strong 'gut health' consumer benefits.
- *Diversify sectors within existing geographical markets*: To-date, the supermarket sector has attracted the lion's share of attention from FMCG companies, as this has been the sector with the greatest volume and value. However, its relative importance is declining and there are other routes to the consumer; for example, there is continued growth in the food service sector. In history, food service has not been well-serviced itself. Products focused at this sector have tended to be the wholesale box and/or larger size versions of products that are merchandised at grocery retail. Further, what was once a fragmented sector and difficult to penetrate is becoming increasingly concentrated and merits the attention to customer requirements that has been characteristic of the grocery retail sector.
- *Explore new distribution channels*: New routes to the consumer are emerging which by-pass the main line grocery sector. The much-discussed Internet shopping, and/or the more generic home delivery sector, offers substantial potential for the future, although even now these routes are being developed by the more proactive supermarket chains, e.g. Tesco.
- *Invest in and support strong primary brands*: Leading manufacturer brands continue to have an important role in grocery retailing. However, this requires continuous investment in R&D to maintain technological superiority over the supermarket 'me-too's', and in promotion, i.e. going over the heads of the retailers to communicate directly about the branded product's benefits with shoppers and consumers. Promotional monies should be focused on brand building, i.e. targeted closely to building brand loyalty, rather than inducing the retailer to carry the brand or merchandising it at a discount, e.g. the potentially damaging 'buy one, get one free' (BOGOF) that serves to confuse the shopper and reduces the overall integrity of the brand. Investing in brands is resource intensive and, by definition, makes it very difficult for smaller-scale companies to compete at the national and, certainly, the international level with the FMCG companies and major retail own-label programmes.
- *Out-compete and/or purchase competitor brands*: In the UK, the PepsiCo-owned Walkers Crisps dominates the crisp/chip market through a combination of product innovation and strong promotion. The purchase of competitor brands has, in history, been a common mechanism for securing market share and consolidating brand strengths. More recently, as manufacturers have rationalised

their product offer and concentrated on 'core strengths', there has been a redistribution of brands between the major players.

- *Produce supermarket own-label products for principal customers (who may be competitors)*: This is a topic which, arguably, is contentious. Some FMCG companies (e.g. Unilever, Procter & Gamble) have a policy of not producing own label products for the major retailers. Others take a more flexible view: H.J. Heinz, for example, produces a greater volume of private label products than it does for its Heinz product range and associated branded products. The issue of manufacturing both private label and nationally branded products is not one of commercial integrity, rather, it is of ensuring that the focus on constant improvement of physical product, image and consumer awareness is not lost on the company's premium, nationally branded products.
- *Focus on markets where the company's brands are or can be number 1*: The research evidence shows unequivocally that companies earn the most attractive margins in markets in which they are the market leader.
- *Provide retailers with the opportunity to earn excellent margins on one's products*: Retailers respond well to manufacturers that sell products that make shoppers/consumers very happy, have high net margin per unit potential for retailers, and are high sales volume items. This combination will result, inexorably, in the manufacturer receiving premium shelf-space which, in turn, should convert into buoyant market share and good profitability.

Overall, the retail sector for grocery products is dynamic and the current market leaders will not necessarily be the market leaders in a decade or so. Manufacturers should seek to develop long-term partnerships with those retailers who will lead the industry in the future. Partnership development, like any investment, is often both human and capital resource intensive. The current industry interest in category management and efficient consumer response (ECR) provides major companies, in particular, with substantial opportunities to gain 'category captain' status, and garner the competitive advantage that such a position can bring.

Small- and medium-scale food manufacturers: surviving and prospering in a competitive world dominated by commercial elephants

In a commercial world dominated by monolithic global retailers and manufacturers, can smaller-scale firms survive and prosper? The answer is 'yes', but it's not going to be a piece of cake!

The research evidence serves to confirm intuition and shows that the closer a company is to its consumers, the clearer are the signals about what are the exact requirements of that consuming group. Further, the fewer links in the chain between manufacturer and consumer, typically, the higher the margin accruing to the manufacturer. In summary, ideal supply chains (surely, demand chain is closer to the mark) are short, fast, transparent and seamless. Yet, so often they are complex, price-driven, confrontational, disjointed and opaque. The former is better than the latter!

The start of a supply (demand) chain for a large or small manufacturer is identifying the precise wants of the shopper and final consumer; the end is confirmation of the total satisfaction of these wants. Yet, too often, for firms of all sizes, the manufacturer relinquishes responsibility once the product is sold to the next link in the chain, i.e. when the manufacturer makes a sale to the retailer and the retailer makes the sale at the check-out. This is not good enough. The satisfaction measure should be post-consumption, culminating in a replete consumer murmuring, 'Marvellous, let's buy another!'.

Irrespective of whether a manufacturer be small- or larger-scale, the requirements of customers are becoming more demanding. In the early 1990s, retailers would seek suppliers who could deliver technical excellence, accurate service levels (delivering what was promised, when was promised and to the specifications promised) and be price competitive. In the first decade of the 21st century, the 'bar' has been raised. Now, customers wish to deal with suppliers, large or small, who have a strategic orientation and, not least, understand exactly what motivates the shopper to buy. The keys to growing the overall product category are innovation in NPD, supply chain management, financing, and effective communication with consumers. Suppliers need to be willing to offer some degree of exclusivity to customers in an increasingly crowded marketplace and can offer the prospect of strong profitability for the customer in exchange for access to scarce retail shelf space.

It is salutary to look at what elements bring success for companies wishing to launch and grow branded food products in the UK. After all, the record shows (ACNielsen, 2001) that nine out of ten new grocery products fail, as measured by their removal from the retail shelves within 12 months of product launch. ACNielsen (2001) identified six factors that are particularly influential in inducing successful and sustained new product launches:

(1) True product innovation has significantly greater success than 'me too' product launches or banal product line extensions ('exciting new raspberry flavour').
(2) New products must deliver the concept promise, e.g. most convenient, best tasting.
(3) Long-term marketing and R&D support are essential for sustained success.
(4) The big firms are more successful at NPD; they are better resourced and better practiced than new and smaller-scale entrants.
(5) The first-to-market with a genuinely new product has a distinct advantage over competitors who arrive later.
(6) Inducing consumer trial is a key to successful launch as there are thousands of products in most retail stores and the manufacturer has to make herculean efforts to bring the product to the attention of the hurried and harassed shopper.

For smaller businesses to survive in the food industry, they must combine the techniques and skills of working with retailers and food service companies in managed supply chains that are dominated by large companies and, most likely, to implement a niche marketing strategy in doing so. If they are successful, this may lead to eventual acquisition by a large company; however, the small business sector has always been a nursery for new ideas. Irrespective of this, the exponentially declining cost of information technology is enabling small firms to act like large firms in their relationships with supply chain partners, and although small business size may suggest resource starvation, it may also indicate nimbleness.

One interesting question is whether companies can graduate from one size category to the next. Perhaps like products, industries have life cycles. Examples of companies graduating from small to very large can be seen in growth-orientated industries such as computer software, fast foods and sportswear. However, within the mature food industry, whilst there are pockets of strong growth in the UK market (e.g. chilled ready meals, ethnic foods), larger-scale companies hover awaiting indications of success from smaller-scale participants as a means of bolstering their own sales momentum and profitability. In food, global firms have come to dominance by different routes. Some have exploited technological break-throughs, such as freezing food or innovative packaging, e.g. Tetrapak. However, most have exploited the economies of mass marketing as a concept to target segments, such as soft drinks, chocolates and spreads. The technologies are often well-established, and competitive advantage is achieved, in particular, through superior management and marketing. With the escalating budgets associated with biotechnological R&D and international promotions raising the costs of entry to the global food industry, the prospects of graduating to the top divisions for smaller-scale firms look particularly challenging.

For the smaller-scale manufacturing firms, dealing with major supermarkets, be it on a national or even a regional basis, can be daunting. Irrespective of how comfortable the commercial relationship may seem, it is like sharing a bed with an elephant; the smaller party can be badly damaged (e.g. through a change in payment policy) without the larger bedmate being aware. Clearly, managing the commercial relationship for the smaller-scale firm involves substantial challenges and the élite supplier must do the following:

- Identify the customer who can add the most value to the product. In the context of the UK, the retail chain which can command the highest and most consistent premium on, say, chilled fresh ethnic foods is one of the 'up-market' high street retailers (e.g. Marks & Spencer [M&S] or Waitrose). These companies have an international reputation in forging close partnerships with their core suppliers. For example, Waitrose management talk in terms of 'nurturing' their valued small-scale suppliers.
- Identify the customer who will be the market leader in the target category in five and ten years' time. Are there existing and/or new companies who will be market leaders in the foreseeable future?
- Identify the most profitable customer, e.g. M&S or Waitrose may be principal customers of a particular supplier, but are they profitable customers?

Supermarket chains can be hard taskmasters and ask a great deal of their suppliers. Are the efforts of serving them worthwhile? Importantly, does the supplier have access to the required financial information to make an assessment of the relative profitability of each of its principal customers, e.g. can and does the supplier undertake activity based costing?

- Manage a balanced portfolio of larger-scale customers. For example, this could involve an exclusive sectoral relationship with one of the major grocery retailers, but include customers from other sectors, such as airline catering, petrol/gas station forecourt convenience retailing and direct home delivery.
- Specifically, the smaller-scale company that is quoted on the stock exchange, needs to consider its position on private or public ownership. In the UK food sector, 60% of listed businesses are capitalised at less than £50 million. Typically, these businesses receive low listings, reflecting poor liquidity in their shares, reduced research coverage by brokers, and a lack of institutional shareholder interest. Indeed, analysts discount the shares of manufacturers of retail private label products, even if these products are in high growth categories (such as chilled ethnic foods), on the basis that the retailer holds all the market power in the partnership. Yet, notwithstanding their modest margins and reliance on the custom of a very few retailers, such specialist private label producers can be strongly cash generative, manufacturing food products that are purchased frequently by a significant proportion of the population. If other sources of capital are available (e.g. venture capital), 'going private' is a *bona fide* option for the medium-scale manufacturer.

If fewer, larger, more globally-orientated supermarket chains with a mass market orientation were the only route to the consumer, then the pessimist might see a bleak outlook for smaller-scale businesses in the UK food industry in the 21st century. But, the optimist may hang a hat on interesting consumer developments in the mature markets of the northern hemisphere. In the face of globalisation of supply sources (e.g. over the past five years, Brazil and Thailand have emerged as key sources for chicken in the UK market), consumers are signalling their increasingly strong interest in local, regional, traditional and seasonal food products that do not suit the large production runs of major manufacturers. Factors driving demand for such products include those that are consumer- and media-led:

- Affected by global events outside their control (e.g. nuclear pollution in former USSR states), consumers seek comfort in local heritage and traditions.
- In countries with an ageing population such as the UK, there is a yearning for traditional and seasonal fare.
- Food is much more than fuel and we use it to make statements about ourselves, e.g. 'I am concerned about environmental pollution caused by long distance food transportation, so I buy local produce'.
- Concern about global food safety may cause us to trust what we know best, i.e. local products.

- Local, regional, traditional food products are 'nice-to-buy' when we are in leisure shopping mode.
- Interest in 'foodie' TV shows and magazines is at a high level, and re-acquainting ourselves with our culinary heritage (real or imagined) is much promoted.
- Finally, expenditure on food is a relatively small proportion of our family budgets. Eating 'something special' is an affordable treat.

However, the 'retro' trend towards regional and traditional food products is trade- and technology-led too, because of the following factors:

- Desperate to differentiate themselves from the competition, major retailers are seeking exclusive access to unique local and regional products.
- There is a renaissance of farmers' markets and food fairs, with craft food products much in evidence.
- Ironically, leading edge initiatives, such as category management, only serve to reinforce the importance of local and speciality foods.
- Restaurants and other food service outlets see incorporating speciality products on their menus as a mechanism for differentiating their offer from mainline retail.
- Saturated global commodity markets force producers to seek higher value market opportunities.
- IT developments which are affordable for small businesses facilitate linkages with larger-scale customers.
- Pressure from customers, lobbyists and competitors is inexorably raising the profile of local and speciality foods.

Perspective should be maintained: the major part of the UK food and beverage market will be serviced by the well-resourced, but slow to respond FMCG 'elephants' who will lumber along with their love–hate relationship with the major retailers. However, smaller-scale firms will have clear opportunities in the high value fresh and freshly-prepared food products sector and in the short production run speciality product sector. The key factor for these companies will be in the identification of the most attractive, i.e. highest value over the long-term, supply chain that each firm can access. Commodity markets are fragmenting and smaller-scale businesses should avoid the declining price-driven majority segments and focus on the segments that are driven by value (and not solely price).

Smaller-scale companies should embrace the third of the Porter generic business strategies, i.e. focus, and thereby gain an intimate knowledge of the specific needs of a very few customers and accord them the highest service level that is consistent with the returns on offer. Nimble, smaller-scale businesses, increasingly sophisticated technologically, will have opportunities to manage food safety risk for major retailers on niche foods, providing them with a full range of premium private label products that are in high growth, destination categories within the store. The disproportionately high profile these products will have in the mind's eye of the shopper (given their relatively modest sales levels) will, paradoxically, bind them closer and closer to their major customers. This proximity will be cosy, but also dangerous, as sharing a bed with an elephant is risky; the pachyderm may flatten you without knowing it. Clearly, choice of a strategic partner will be acutely important to the smaller-scale business, as the number of partners it can service will be constrained and, therefore, its customer portfolio will, by definition, be unbalanced.

Increasingly in the future, the larger players will wish to have exclusive suppliers, as in saturated grocery markets characteristic of northern Europe, and major vendors will wish to differentiate their product offers from their direct competitors, with supply chain-based competition becoming pervasive. It will be in the interests of major customers to keep their coterie of suppliers in business for the long term. For many smaller-scale businesses, the evolving new world of commerce will require them to undergo a cultural *volte face*. Their customers, retail, food service and others, will seek suppliers that understand the dynamics of the market and can assist them directly in developing strategy to build category sales and market share. This will require a profound knowledge of shopper and consumer behaviour, and necessitate unfamiliar investment in market research. Within a specific chain, members will seek to encourage 'chain learning', i.e. to eschew their confrontational past and maximise the chain's competitive advantage through self-perpetuating improvements. Members will share strong mutual commercial interests with their vertically-aligned suppliers and customers.

A final word

Big or small, food manufactures will face considerable challenges as the UK, European and global food industries evolve and intertwine over the years ahead.

In the UK and elsewhere, market power is polarising in the food industries of the world. At one end of the chain, fewer and larger international life science companies are emerging that seek proprietary ownership of intellectual property relating to the genetic building blocks of food products (and of life itself). They see the future offering substantial opportunity for them in owning the rights to key attributes, much valued by consumers, that can be incorporated into food products. Such attributes might be health benefits or anti-ageing remedies. At the other end of the chain are global retailers who are harnessing IT to make sense of our shopping behaviour from week to week. Twelve million shoppers hold Tesco 'Club' cards and the information collected each time a card is swiped at the supermarket checkout is proprietary and very valuable. The concern is that all other food industry participants in between these monoliths will get squeezed, unless they have their own proprietary product and/or service. For a Unilever or a Nestlé, their brands are well-embedded in the consumer's psyche and are proprietary; they reflect years of investment. This is the challenge for the smaller-scale and resource-constrained manufacturing company that needs to work out how to add value for its customers and in a manner that cannot be immediately copied (product or service) by its competitors.

Study questions

1. In a commercial world increasingly dominated by large-scale, even global, retailers, how can larger- and smaller-scale food manufacturers survive and prosper?
2. Looking towards the future, apart from traditional supermarkets, what routes (marketing channels) will food manufacturers use to reach consumers in the next decade of the 21st century?

References

ACNielsen (2001) Retail trade. *Checkout*, December 2001. Cumulus Business Media, Cheam.

Bolling, C. & Somwaru, A. (2001) *U.S. Food Companies Access Foreign Markets Through Direct Investment*, Economic Research Service, USDA, Washington DC.

DTI (2002) *The 2002 R&D Scorecard: Commentary and Analysis*, Department of Trade and Industry, London.

Gilpin, J. & Traill, B. (1995) Small and medium food manufacturing enterprises in the EU: the case of the United Kingdom, *Discussion Paper No. 23, Structural Change in the European Food Industries*. EU AAIR programme, European Commission, Luxembourg.

Henderson, D.C., Handy, C.R. & Neffs, S.A. (eds) (1996) Globalisation of the processed food market, *Agricultural Economics and Rural Sociology report no. 742*, Economic Research Service, USDA, Washington DC.

Hughes, D. & Ray, D. (1999) *Developments in the Global Food Industry: A 21st Century View*, Wye College Press, London.

Information Resources Ltd (2002) Viewpoint, *The Grocer*, 14 December 2002. William Reed Publishing.

ONS (2002) *Annual Business Inquiry*, Office of National Statistics, Newport (www.statistics.gov.uk/abi).

Porter, M. (1980) *Competitive Strategy*, Macmillan, New York.

Porter, M. (1985) *Competitive Advantage*, Macmillan, New York.

Rabobank International (2002) *Food and Agribusiness Research*. Utrecht, The Netherlands.

Taylor Nelson Sofres (2002) *The Grocer*, 16 November 2002.

USDA (2002) Eurofood. In *Global Agriculture Information Network (GAIN) Report*, p. 14. United States Department of Agriculture, Washington DC.

Chapter 8

Food Retailing, Wholesaling and Catering

J. Dawson

Objectives

(1) To review the scope and size of the three sectors of retailing, wholesaling and catering.
(2) To explain the major structural changes in these sectors, particularly the issues of market concentration, format development, branding, and customer responsiveness.
(3) To suggest how the sectors may develop over the short term.

Scope and size of food retailing, wholesaling and catering

Retailing is the stage of the chain where firms interact with final consumers as customers. Wholesaling has a business based customer base and catering has both businesses and final consumers as customers. This difference in customer base has a major impact on the strategies and operations of the firms in the three sectors. For retailers, major issues are communication with final consumers through a store network and through marketing initiatives such as branding. The market is large and spatially disaggregated. For the wholesaler the number of customers is much smaller and the dominant need of these customers is the availability of product. The catering sector is more complex because it includes a network of establishments that serve the final customer, through restaurants and bars, and also provides a service into other business through food service provision to institutions such as hospitals and schools, and also to private sector firms who outsource their catering needs. Given this difference of the customer bases for the sectors, it is not unexpected that issues of size, strategy and structure are somewhat different.

The firms involved in the retailing and wholesaling of food often have some ranges of non-food items that are integral to their business. Thus many of the firms include ranges of groceries, including for example cleaning and laundry items, health and beauty items, and even basic clothing ranges. As a consequence data on total market sizes and market shares are notoriously difficult to determine. Even within the food ranges there are important differences in the categories of product depending on the temperature regime needed to maintain the food – frozen, chilled, ambient and hot. Food versus non-food issues and temperature regime have important implications on company operations, not least in the inventory management systems needed to maintain the variety of products expected by the various customers.

Retailing

The total sales in the UK retail food and grocery sector in 2000 amounted to approximately £100 billion. The Office of National Statistics (ONS, 2002) in the *Annual Business Inquiry* divides the sector into:

- non-specialised stores with food, beverages or tobacco predominant – sales of £85.1 billion
- stores specialised in food, beverage and tobacco – sales of £12.4 billion.

In addition there are food sales incidental to the ranges in other types of retail outlets. The Institute for Grocery Distribution (IGD, 2002a) calculates that the UK retail grocery sector had sales of £99.8 billion in 2000. Figure 8.1 shows the growth of the sector at current prices through the 1990s with forecasts to 2004. The trend in the year-on-year change in market size through the 1990s is of a decreasing rate of growth. Typical annual growth rates have fallen from 6–7% at the start of the decade to 2–3% by 2001. The inflation adjusted growth in the early years of the current century is close to zero.

Within the food retail sector, there are some 80 000 businesses operating approximately 100 000 shops. The number of specialist food businesses, and shops,

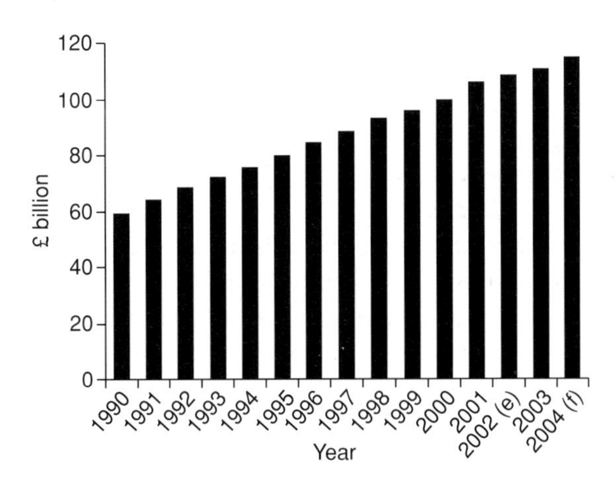

Figure 8.1 Size of the retail grocery market in the UK (e, estimate; f, forecast) (Source: IGD, 2002a).

Table 8.1 Food retailing in the UK (*Annual Business Inquiry*, Office of National Statistics).

Year	Number of enterprises		Number of establishments		Percentage gross value added[a]	Stock turn[b]
	Non-specialised stores with food predominant	Specialist food	Non-specialised stores with food predominant	Specialist food		
1996	22 596	59 476			16.3	21.0
1997	24 164	54 999	39 025	69 852	18.8	20.6
1998	24 724	50 435	41 311	65 098	18.7	21.3
1999	25 864	47 307	42 951	60 476	19.6	19.4
2000	26 562	44 021	42 388	56 913	19.8	21.4

[a] Sales less total value of goods materials and services purchased, as percentage of sales.
[b] Sales: average start and end year stock.

Table 8.2 Market share of grocery sector in December of given year (source: Institute of Grocery Distribution).

	1990	1992	1994	1996	1998	1999	2000	2001
Tesco	9.7	10.1	11.4	14.2[b]	15.2	15.6	16.2	16.8
Sainsbury	11.0	11.9	12.3	12.2	12.2	11.8	11.5	11.7
Asda	6.8	6.3	6.7	7.8	8.4	9.1	9.5	10.0
Safeway	7.1	7.3	7.6	7.6	7.6	7.4	7.5	7.3
Somerfield[a]	7.7	8.1	8.4	8.1	6.9	6.1	5.0	4.6
Morrisons	1.4	1.7	2.2	2.5	2.6	3.0	3.3	3.4
Marks & Spencer	3.4	3.0	3.1	3.1	2.9	2.9	2.8	2.7
Waitrose	1.7	1.7	1.6	1.8	1.8	1.9	2.1	2.1
Iceland	1.2	1.5	1.7	1.7	1.6	1.7	1.7	1.6
Co-operatives	14	N/A	N/A	N/A	N/A	N/A	6	N/A
Independents	10	N/A	N/A	N/A	N/A	N/A	6	N/A

[a] Includes Kwik Save acquired in 1997.
[b] Purchase of William Low in September 1994.
N/A, data not available.

fell substantially in the late 1990s, with a decrease of 11 000 businesses and 13 000 shops over the last four years of the decade: the figures are shown in Table 8.1. Although the total number of businesses is substantial, a relatively small number of large firms account for a large proportion of total sales. The IGD estimates that the nine largest firms accounted for 60% of sales in December 2001, having increased from 50% by these same firms in 1990. Table 8.2 indicates the market shares of these firms between 1990 and 2001: Tesco and Asda have grown their market share by the largest amount; co-operatives and independent small firms have lost market share.

Percentage gross value added is close to 20% of sales for the sector (Table 8.1). Gross value added is calculated as sales less the costs of goods and services purchased. It represents the amount available for various business costs, including employee costs and internal costs associated with the various business functions, for example marketing, logistics and store development. Clearly profit is also included within this gross amount. The trend in the last few years is for the percentage gross value added to increase (Moir, 1990; Burt & Sparks, 1997), mainly due to two factors: a policy of outsourcing some functions so that they become bought services rather than internally provided functions and a general shift in the food chain towards retailers as the location of added value (Dawson & Shaw, 1990; Ogbonna & Wilkinson, 1998).

A second important indicator is stock turn. Whilst there is small variation from year to year, the average figure for the sector is approximately 20, but this varies considerably by type of retailer and by product area; for example, for fresh food specialists, e.g. fishmongers and bakers, stock turn is much higher, but for frozen food retailers it is much lower. It is important to remember that the food retail sector is not homogeneous but contains very great variety of firms and operations. In this respect, average figures can sometimes be misleading and have to be interpreted carefully.

Wholesaling

The food wholesale sector includes businesses that are selling to processors and manufacturers within the chain, and those that are selling to retailers, caterers and institutions at the end of the food chain. In 2000, the total sales of all UK wholesalers of food products were £61.4 billion through slightly fewer than 15 000 enterprises. Table 8.3 shows that the number of enterprises is falling and the trend in total sales is one of a very low increase. Wholesaling in intermediate activities is much larger than at the end of the chain and is decreasing in its level of activity. A function of wholesalers is to hold stock, and this can be seen in the relatively low stock turn shown in Table 8.3.

The sales of the grocery wholesale sector to retailers, caterers, etc. at the end of the chain in 2000 were £15.3 billion, representing a 50% increase since 1990. There are two very different parts to this section of the wholesale sector: cash and carry, and delivered trade. Generally, the customers in cash and carry are the small independent retailers, caterers and hotels who visit the wholesaler to collect the

Table 8.3 Food wholesaling in the UK (*Annual Business Inquiry*, Office of National Statistics).

Year	Number of enterprises	Total turnover (£ million)	Gross value added[a] (%)	Stock turn[b]
1996	15 422	61 653	9.0	18.3
1997	15 818	64 035	11.8	15.8
1998	15 446	60 815	9.6	16.9
1999	15 134	60 784	11.6	17.7
2000	14 826	61 403	9.4	19.1

[a] Sales less total value of goods materials and services purchased, as percentage of sales.
[b] Sales: average start and end year stock.

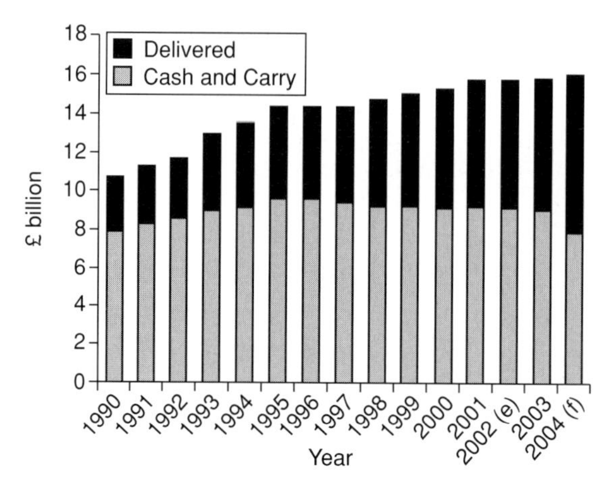

Figure 8.2 Wholesaler grocery sales to retailers, caterers, etc. at the end of the food chain in the UK (e, estimate; f, forecast) (Source: IGD, 2002b).

products that they sell at retail or use within their business. It is not surprising, therefore, that the cash and carry sector, although the larger of the two sub-sectors, has declined since the mid-1990s. Figure 8.2 shows the changing balance between the two sub-sectors (IGD, 2002b). As with retailing, a small number of firms dominate the wholesale sector. These firms are often linked to voluntary chains of retailers, notably Spar and Mace, that provide a solid base of customers for the wholesaler.

Catering

In contrast to the other sectors, in catering the number of enterprises is very large and is increasing. This is particularly evident in the restaurant sub-sector where the number of enterprises increased by 6000 between 1996 and 2000. The sector is often divided into a 'profit sector' that comprises restaurants (including those in hotels), fast-food outlets, bars and public houses, and the 'cost sector' comprising meals provided in the workplace, education, health care and welfare services. It is estimated that there are approximately 175 000 outlets in the profit sector and 85 000 in the cost sector.

Table 8.4 illustrates the general position in recent years in respect of enterprises rather than outlets. Gross value added is much greater in this sector than in retailing or wholesaling, but labour costs account for a relatively large share of this margin. As would be expected, stock turn is high with low levels of stock held compared with sales.

The changing structure of the sectors

The retail, wholesale and catering sectors were very traditional sectors of the UK economy for many years. From the mid-1980s onwards they started to undergo a

Table 8.4 Restaurants,[a] bars, canteens and catering in the UK (*Annual Business Inquiry*, Office of National Statistics).

Year	Number of enterprises	Total turnover (£ million)	Gross value added[b] (%)	Stock turn[c]
1996	95 326	26 697	40.4	46.8
1997	95 401	29 625	42.4	49.7
1998	96 986	31 407	42.4	44.9
1999	99 284	33 357	42.3	44.8
2000	101 701	35 951	42.5	45.1

[a] Excludes restaurants in hotels.
[b] Sales less total value of goods materials and services purchased, as percentage of sales.
[c] Sales: average start and end year stock.

transition. It has been suggested that these sectors have been the subject of three major and fundamental macro-innovations in the last 50 years. Initially there was the adoption of self-service: starting in the late 1940s, this was followed 25–30 years later by the acceptance of marketing as the dominant business paradigm. Most recently, the convergence of information and communication technologies is the latest fundamental macro-innovation that is changing the structure of the sector.

Self-service had a major impact on retailing, with the customer having direct access to the products, larger stores becoming possible, and employees having very different functions to perform. Wholesaling adopted self-service a little later than retailers with the introduction of cash and carry operations, and the self-service cafeteria diffused through the catering sector (Hunt, 1983). The second macro-innovation, that of the acceptance of marketing, resulted in designing delivery mechanisms to suit the needs of specific types of consumers and specific consumer demands. In retailing, targeted store formats were the result, with the emergence of superstores, convenience stores, limited range discount stores, etc., all operated as means to satisfy a specific, rather than generalised, aspect of demand. In meeting the needs of specific consumer targets there was a need to know exactly what products had been bought by consumers, and so bar-codes became widely accepted as a means of identification of products. In catering, the emergence of themed restaurants and bars was a response to this acceptance of marketing as the dominant paradigm. The convergence of information and communication technologies is presently enabling the successful management of very large networks of stores and catering outlets, an impossible task before the latest macro-innovation, and the effective introduction of remote shopping and remote wholesaling using the Internet.

With each of these three major innovations, there has been an associated development of new sales formats. A sales format is a sales outlet that has been standardised to meet the needs of a particular consumer segment. Individual firms adjust the format to represent their own brand and so create a formula. Thus a convenience store is a format, with Tesco Express, Shell Select, Sainsbury's Local, etc. being branded formulae. Each of the three macro-innovations results in a proliferation of

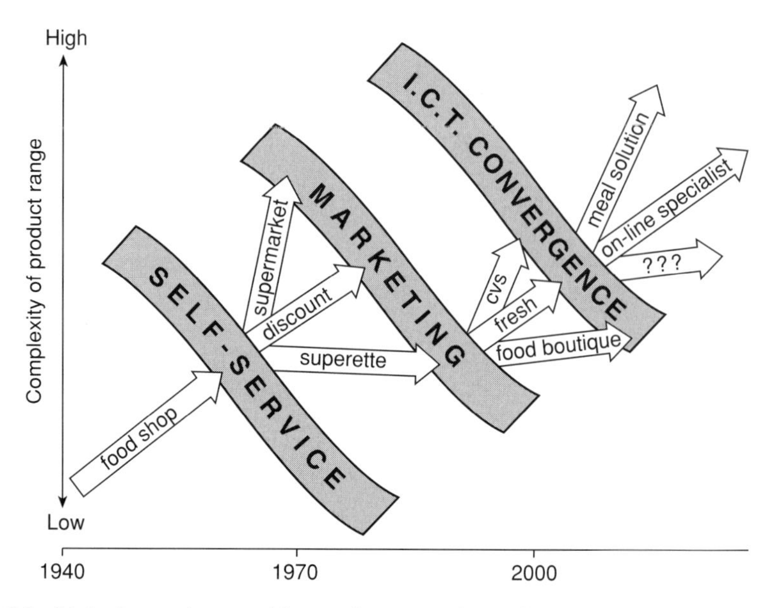

Figure 8.3 Major innovations and format fragmentation in food retailing.

formats. Figure 8.3 illustrates this, with examples drawn from small retail formats in the food sector. Thus, with self-service a number of formats were generated. The small superette format then fragmented as a consequence of the acceptance of marketing. Each of these formats is then fragmenting again as we currently pass through the innovation phase associated with the convergence of information communication technologies (ICT).

Market concentration

In all three sectors considered in this chapter, there is a continuing trend to market concentration (Akehurst, 1983; Davies *et al.*, 1985). In retailing, this trend is apparent in countries other than the UK and in other parts of the retail trade. The reasons for this are as follows:

- There are considerable cost economies of organisational scale within food retailing, wholesaling and catering. It should be noted that these scale economies are greatest at firm level rather than establishment level. They arise in several areas, but most notably in purchasing. In retailing, the cost of goods and services purchased amounts to over 80% of the sales (Table 8.1) and so a reduction in costs of purchasing can affect profitability in a significant way. Larger firms have lower unit costs of purchasing because of savings on search costs and, most importantly, stronger negotiating positions, with suppliers giving volume discounts on price (Dobson *et al.*, 1998).
- There are scale economies to be gained by the replication of store and catering formulae, with the design and system cost being spread across many establishments.

- Scale economies also accrue in respect of marketing costs, with advertising and promotional costs being responsive to scale. Larger firms can access potential customers at a lower unit cost than smaller firms.
- Larger firms have more resources such that they can develop customer communication systems, for example loyalty cards, at lower unit costs than small firms with fewer customers.

Retailing and wholesaling have become more concentrated in the past decade, with most of the larger firms growing at a faster rate than the market. In the UK cash and carry sector of wholesaling, four firms accounted for 49% of the market in 2000, with Booker's share being approximately 35%. This consolidation can also be illustrated by the acquisition activity of Palmer Harvey McLane, a long established delivered trade wholesaler with an estimated 45% of the delivered grocery wholesale market: in 1995 they acquired Snowking Frozen Foods, a frozen food wholesaler; in 1998 they acquired Winerite, a specialist beer, wine and spirit wholesaler based in Yorkshire; in 1999 they acquired the delivered wholesale activity of Booker which served the Mace stores in England and Wales; in 2000 they purchased a van-sales wholesale operation, Snacksdirect, from KP McVities. In the catering service, the two firms 3663 and Brake Bros have become dominant through a series of acquisitions. The catering sector remains highly fragmented, although a small number of large firms have emerged in recent years and these have grown strongly. The result is that the level of market concentration in catering is increasing, although from a low level.

Formats and formulae

An important factor that has affected the structure of the sectors has been a strategic move by firms to develop the concept of formats and formulae. This strategy has been most evident in retailing, but is now widely used also in catering, and is becoming more important in wholesaling. The idea of format requires the identification of customer segments, each of which can be targeted by a particular combination of the marketing mix factors. Thus in retailing, some customers seek low prices as their over-riding means to satisfaction, and discount supermarkets have been developed as a format in which different firms have slightly different approaches, so creating the formulae Aldi, or Lidl, or Netto, or Kwik Save. Other food retail formats are grocery superstores, hypermarkets, convenience stores, supermarkets, food boutiques, category specialist, Internet sites, vending machines, etc. Each format has distinctive operating characteristics. The formula results from the branding of a format by a firm.

The range of variables that can be used to design a format and the branded formula is considerable (Coopers & Lybrand, 1996). The traditional marketing mix variables are extended considerably. Figure 8.4 illustrates the range of variables that is available to be managed and also shows how these variables score on a typical discount supermarket.

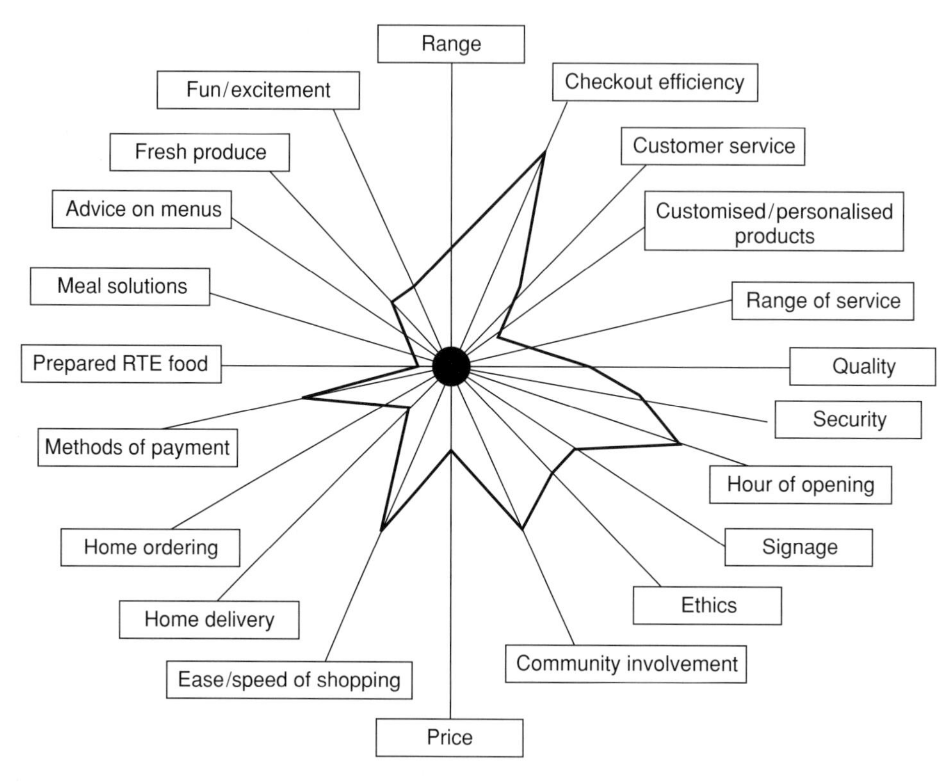

Figure 8.4 Variables in format design applied to a limited range discount store: low values are in the centre and high values on the periphery of the diagram (Source: Coopers & Lybrand, 1996).

Within all the formats, price is an important aspect of the marketing design. The relative importance of price varies from format to format. In discount supermarkets, price is the dominant factor, with typical prices at least 5% below market averages (Burt & Sparks, 1994). The discount supermarket format generally has 1000–3500 product lines (stock keeping units – SKU) in a store of between 800 and 1300 sq. metres. The smaller stores with fewer SKUs (e.g. Aldi formula) are generally termed 'hard' discounters, whilst the larger stores with bigger range (e.g. Kwik Save) are termed 'soft'. In both cases, the business model is one of focusing the scale of buying onto relatively few products such that low unit prices of purchase is obtained with this being passed on in low retail prices.

In addition, store designs emphasise the low cost nature of the operation. For success in this model there is a need to have a large number of stores such that buying scale economies can be achieved. The major operators are international and combine their buying power from stores in several countries. Three hard discount retailers entered the UK in the early 1990s. Before that Kwik Save was the only major operator, with others either having left the market or having changed their strategies to the operation of different formats, e.g. Tesco. The German company

Table 8.5 Major discount supermarket retailers in the UK (data from companies).

Company	Number of stores			Average store size in 2001	Percentage of shoppers in		Gross sales (£ per sq. metre per week)	
	1998	2000	2002	(sq. metres)	1 & 2 person households	Over 45	1999–2000	2001–2002
Kwik Save	873	808	710	797	58	62	68.9	63.8
Lidl	126	203	345	935	56	66	73.7	67.8
Aldi	198	236	263	790	51	56	110.7	110.4
Netto	117	120	126	630	49	46	95.6	89.4

Aldi entered the UK in 1990, the Danish firm Netto entered in 1991 and the German company Lidl entered in 1994.

All three have built up a substantial presence since entry. Their typical requirement for a new store is a catchment population of 10 000, a town centre or edge of town location, 0.3–0.5 hectare of land that will allow a store of about 1200 sq. metres, a high visibility site at a road junction with access off the principal road. Table 8.5 shows the extent of the networks of these four major firms using this format. The individual formulae differ slightly, with Lidl, for example, targeting older smaller families, and Netto aiming at younger customers with large households. Discount supermarkets account for about 6% of sector sales, but this has fallen slightly in recent years as other formats have become more aggressive with their pricing policies.

The dominant retail format is the grocery superstore (Wrigley 1992, 1994). This is usually considered to be a store of over 2500 sq. metres with a full range of food items in all temperature regimes and with a limited range of non-food items. In 2002, there were an estimated 1240 of these stores, accounting for almost 40% of sector sales. The four major operators, Tesco (329), Sainsbury (302), Asda (239) and Safeway (186), together account for 85% of the total. The strength of the format is the range of products available and the opportunity for customers to purchase all their food and grocery requirements in the one store.

An important development in respect of formats is the move by the major firms to operate multiple formats and so target different consumer segments with different formats (Coca Cola, 2001); Table 8.6 illustrates the position. The four major companies are all pursuing this strategy which has progressed furthest with Tesco. Tesco, Sainsbury and Safeway are all developing networks of small convenience stores and very large hypermarkets to complement their superstore and supermarket portfolios. Asda is developing hypermarkets (termed supercentres) to add to their network of superstores.

The strategic use of formats and formulae is also apparent in the catering sector. Whilst the traditional fast-food burger formats have been present for some time, there are a variety of other formats that have become more formalised over the last 10 years. Sandwich shops, coffee shops, family restaurants, etc. have become more

Table 8.6 Multi-format strategies of major food retailers in the UK (company data).

Company	Trading formulae	Number of stores in 2002	Average size of store (sq. metres)
Asda	Asda	245	4200
	Asda Supercentre	4	9600
Safeway	Safeway Megastore	5	6000
	Safeway[a]	475	2100
	Safeway–BP	55	200
Sainsbury	Savacentre	11	8200
	Sainsbury	410	3100
	Sainsbury Central	8	1100
	Sainsbury Local	18	280
Tesco	Tesco Extra	53	6800
	Tesco	453	3000
	Tesco Metro	142	1200
	Tesco Express	65	200

[a] Safeway have introduced a strategy of reformatting these to either Safeway Superstores or Safeway Supermarkets. By late 2002, 140 stores had been reformatted into 65 Superstores and 75 Supermarkets.

consciously designed as sales formats for food purchasing, and within the bar sector, wine bars and theme pubs have become commonplace, with again different formulae, and sometimes several formulae, being developed by specific firms. The bars and restaurant formulae have taken on many of the characteristics of brands.

Brands and branding

Branding in the food distribution sectors is strong (Burt, 2000); it operates at three levels. There is branding of the firm, branding of the operational unit (supermarket, cash and carry warehouse, fast food restaurant, etc.) and branding of the item sold to the customer.

Branding of the firm has become increasingly important as a means of communication with both the customers and the city financial institutions that often have substantial investments in the larger firms. Thus, Tesco in retailing, Makro in wholesaling, Brake Bros in catering service, and McDonalds in catering, for example, have become corporate brands with brand identities recognisable by their customers and investors.

Branding the format emerged during the 1990s as a major strategy of large firms. As multiple formats are developed, so each carries the corporate brand name and a formula brand, for example Tesco Extra, Sainsbury Local, etc. The logic behind this branding exercise is a belief that in respect of marketing activity the 'store' is the 'product' of the retailer (Dawson, 1995). The same concept is applied in the catering sector with restaurant chains or themed bars. In effect, the operating establishment is a designed entity that is 'consumed' by the customer who collects a combination of items to be bought via a shopping trolley of goods or a meal off the menu. The establishment (shop, bar, etc.) also becomes a medium for communication with the customer. This is achieved by designing the sensory stimuli (visual, aural) within the designed space. It is a short step from considering the shop, bar, etc. as designed space and the branding of it. This type of branding is designed to build loyalty with the customer and to communicate brand values, such as high quality, convenience, fun and sophistication, depending on the target market being addressed.

The third area of branding activity is the branding of the items that are purchased by the customer. Moves in this direction by food distributors have been underway steadily for 30 years or more, but they have increased in importance and extent since the late 1980s. Retailer branding is also termed as own label, distributor brands and own brand, but this mix of names can be confusing because it is not only retailers, amongst distributors, who are involved in the branding of individual SKUs. Wholesalers develop brands and apply them to items that they sell to retailers. Caterers brand items they sell in both institutional catering outlets and in outlets accessible to the wider public. For example, 3663 have three brands that are used on items that they supply to caterers: 3663 Smart Choice as the main brand, Coronet as a portion pack brand, and Springbourne as the brand for bottled water. The presence of item branding, nonetheless, is most widely apparent as an activity of retailers.

The share of retailer branded items in the market varies considerably by category. It is high, for example, in chilled ready meals and dry pasta at over 70%, and low in pet foods at less than 15%. There is also considerable variation by firm, with higher levels in Sainsbury and Tesco with a share of between 55% and 60% compared with Safeway and Somerfield where the share is closer to 50%. Figure 8.5 shows the share since 1980 of the dry groceries and toiletries sector accounted for by retailer branded items. From a little over 20% in 1980, the share peaked at

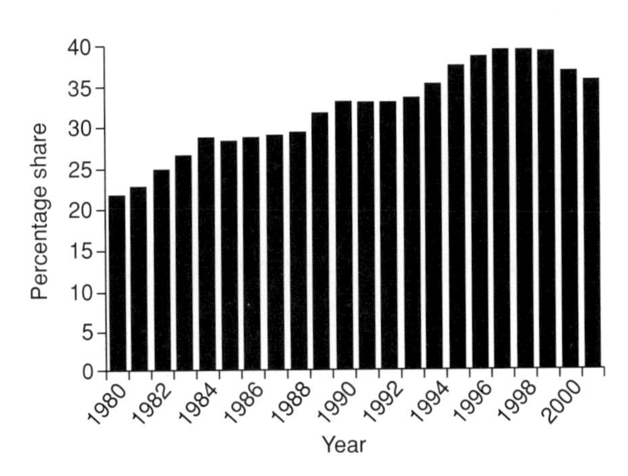

Figure 8.5 Percentage share of retailer branded items in packaged groceries and toiletries category (Source: ACNielsen and Taylor Nelson Sofrés).

almost 40% in the late 1990s and has since fallen back slightly. In some categories the share is still growing, particularly in chilled foods areas. In organic categories the share is growing, but from a low base as the category has only been developed relatively recently.

The reasons for retailers expanding their item brand presence are threefold.

(1) *Store/firm image and customer loyalty*. Good value items enhance the image and build loyalty to firm, store and item range. The retailer's image is carried into the household and is a constant reinforcement throughout consumption. Consumers assume that retail branded items are made by major manufacturers.
(2) *Competitive advantage*. Retailer branded items are unique and provide a basis for competitive differentiation. Because the retailer controls the range of items in a category, the items can be positioned throughout the range, giving better control by the retailer. Promotional activity can be managed more effectively by the retailer.
(3) *Profitability*. Margins on retailer brands are 5–15% higher than on manufacturer brands, so allowing both lower prices and more profit. Tighter stock control is possible on retail brand items and results in increased profitability.

There is a steady increase in the level of sophistication of the retailer branding of items. From being applied to relatively low quality and low price generic commodities, the retail branding activity has moved through a number of phases, with increased marketing support at each phase. Four main types of retail brand have been suggested and are shown in Table 8.7. Each type has distinctive characteristics in terms of marketing and sourcing. Although all four types can be seen today, there has been an evolution over 30 years of the dominant type. In the 1970s, type 1 brands were much in evidence and by the 1990s type 3 were the dominant form with considerable growth in type 4 brands. The unresolved question is, assuming the types reflect stages of brand development, what will be type 5?

The increased sophistication of retailer branding has resulted in multiple positioning of retail brands within the one firm. There has been development, therefore, of retailer brand items of higher value and price than the standard retail brand, and also ranges of discount brands of quality and price below the main brand. The major retail firms have all followed this policy. For example Asda has the following retail brand ranges:

- Asda core brand with over 7000 items in the range, positioned to match the major manufacturer brands and other retailers' core brands. The brand encompasses food and non-food.
- Asda Smartprice is positioned as a low price brand with claims that competitors' items will not be priced below this. There are approximately 650 items in this range.
- Asda Extra Special is the premium brand with about 250 items in mid-2002 and a further 200 in development. Packaging is designed to support the claim of high quality food.

Table 8.7 Types of retail brands (based on Laaksonen & Reynolds, 1994).

Factor	Type 1	Type 2	Type 3	Type 4
Branding form	Generic	Unsupported retail brand	Supported retail brand	Segmented and sub-brands
Strategy	Generic	Low price copy	Copy of major brands	Added value
Objective	Improve margin	(a) Improve margin (b) Reduce manufacturer power	(a) Improve margin (b) Reduce power of manufacturer (c) Extend assortment (d) Build image	(a) Improve margin (b) Extend assortment (c) Build image (d) Create differentiation (e) Improve customer loyalty
Product	Commodities and basic products	Basic products with large volume sales	Major sale items	(a) Niche products (b) Special category product
Technology	Simple process	Copy of market leader	Comparable to main brand	New technology and processes
Quality	Low	Medium and below market leader	Comparable to market leader	Same or better than market leader
Price position	20% below main brand	10–15% below main brand	2–5% below main brand	Premium
Consumer motivation	Low price	Low price	Value for money	Different product
Supplier	Unspecialised	Specialised and has own brands	Specialised in retail brands	International specialist in retail brands

- Asda Good for You is a range of about 300 items, emphasising healthy eating, with low fat and low salt content.
- Asda Organic had only 100 items in its range in mid-2002, but the range is growing quickly in response to growth in the market.

These five positions for the retail brand items are similar to the positioning used in other firms, although the balance of the ranges varies considerably. Tesco have substantial presence in premium (Tesco Finest) position with almost 1000 lines by late 2002. Sainsbury have expanded their healthy eating positioned range with a sub-brand of Blue Parrot Café providing meals for children. Safeway have ranges in all the main positions, but with a relatively small number of items in the non-core ranges compared with the other major firms.

Branding of firm, format and item has been an area of considerable innovation by food retailers since the mid-1980s. A number of the initiatives of retailers are being copied by wholesalers providing for the catering trade and by firms in the catering sector, for example in the sandwich ranges of Pret à Manger.

Responsiveness to customers

A fourth important feature of change in the three sectors has been the increase in the responsiveness of firms to the changes taking place in consumer behaviour and consumption. The competition for customers by retailers and caterers, who now see each other as direct competitors, for what has been termed 'share of stomach', is apparent in the creation of formats/formulae, brands and a variety of loyalty schemes.

In this competitive process, attempts are made to gain a better understanding of the needs of consumers and in particular to segment consumers in new ways. There is a widespread belief that consumers are becoming more knowledgeable about the activities of food distributors and at the same time consumer needs are becoming more fragmented in quality and in time (Sparks, 1993). The consequence is that consumers not only seek a greater variety of items, being less willing to purchase and consume the same item repeatedly, but also have very different patterns of purchasing behaviour at different times of the day or week. The long established bases of segmentation associated with age, income, social class, etc. have given way to new bases of segmentation linked to lifestyle and space–time related behaviour.

Illustrative of this change is the growth of the convenience sector in both catering and retailing. The willingness of people to purchase coffee 'to go' on their way to work is evident, but there is also a desire to read a book or newspaper in an easy chair while drinking coffee later in the day. This fragmentation is seen also in the snack food market and in the grocery shopping market. Sainsbury, for example, have identified six types of shopping activity associated with the use of their Local formula of the convenience store format:

- 'Grab 'n Go' when food is needed for immediate consumption, typically within 5 minutes of purchase.

- 'Convenience' when key category items include sandwiches at lunchtime and ready meals in the evening.
- 'Perishable top-up' when a planned purchase of perishable items between main shopping trips is taking place, including fruit, vegetables, meat, bread and dairy items.
- 'Impulse' that is a spur of the moment purchase, perhaps as a self-treat.
- 'Habitual' that is a regular, often daily purchase, of items, e.g. newspaper or cigarettes.
- 'Distress' that is a purchase of an essential item(s) that has been used up but is needed before the next main trip, e.g. toothpaste, kitchen towels.

The mix of different convenience behaviours varies with the location of the store; for example, a store in a suburban housing area has a different mix from one at a railway station. The different mix may also result in changes in the store layout to accommodate a dominant customer group, with a 'Grab 'n Go' store having an open easily accessible layout and additional mobile checkouts for peak times.

The responsiveness to consumer needs requires a detailed understanding of the needs and purchase behaviours of consumers. This has resulted in a variety of new methods of collecting information on consumers, including store based focus groups and loyalty cards. The use of data from loyalty cards is now integral to design processes for store development and to in-store merchandising decisions.

Potential future patterns of development

A strong and continuing trend across all parts of food distribution considered in this chapter is that of continuing concentration of the market. In wholesaling, a small number of large firms increasingly dominate the various market segments. In catering, again, there is concentration in the various parts of the market, although in this sub-sector, as in retailing, there are still many small micro-firms. The birth and death rates of these micro-firms is high and their overall market share is decreasing, but nonetheless they are a notable feature of market structure and are unlikely to disappear as a part of the market in the medium term. They provide for niche customer needs either in terms of specialist food items, specialist location, specialist lifestyle or specialist time-frame. The dynamism of the sector means that the birth of micro-firms will continue, with a stream of new ideas to respond to specialist consumer needs, whether for a Moroccan restaurant, a specialist honey shop, a farm-shop, organic red-meat wholesaler, skiers' bar, etc. Within overall market structures, however, these micro-firms provide welcome diversity, but the majority of distribution activity is controlled by ever fewer large firms.

A second aspect influencing the pattern of future development is the extent to which governmental policies, particularly in the areas of land use planning, competition policy and food standards legislation will affect operating costs in distribution (Clarke, 2000; Clarke *et al.*, 2002). A series of policies enunciated in the UK in Planning Policy Guidance notes have established, over many years, the governmental position on the location of large new food stores and on large retail centres that include bar and restaurant formats in addition to retailing. These policies have varied in the degree to which they introduce constraints on new development, particularly in out-of-town and edge-of-town locations (DETR, 1998). Through the late 1990s, the constraints in the UK were increased, with requirements to undertake evaluations of alternative in-town sites, provide proof of the economic need for new stores, and undertake environmental impact assessments of proposed developments. In the next few years, it seems that these constraints are more likely to be increased than to be relaxed. This will affect future patterns for the sector which will be required to develop formats that can be accommodated within towns or that do not need planning permission; for example, Internet facilitated formats.

Competition policy in the UK also will influence the sector over the next five years by limiting the extent of merger activity and potentially affecting costs in the chain by intervening in relationships between distributors and their suppliers. The major report on supermarkets that was undertaken by the Competition Commission (2000) explored many of the practices in the chain and proposed that codes of practice be established to regulate relationships. Such intervention may change cost structures and may affect industry structure by protecting, to some extent, the smaller firms. The conflicts between land-use and competition policies arise when local markets are considered, with land-use policies seeking to limit new stores and so reduce competition, whilst competition policy is attempting to increase the level of competition (Guy, 1996; Wrigley, 2001). Thirdly, food standard policies will have an increasing impact on distributors with the need to adjust formats or create new formats to accommodate the various requirements of the UK Food Standards Agency.

Of considerable importance in generating future patterns are the likely trends and fashions in managerial concepts. The differences in managerial expertise amongst major firms are an important factor in profitability (Seth & Randall, 1999; Wileman & Jary, 1997). In retailing, the four major firms each have individual corporate strategies and these have generated different patterns of growth in recent years. Figure 8.6 shows profitability plotted against market share. Whilst Tesco sought market share rather than margin, Sainsbury had initially sought margin but had to reduce this to maintain market share. Asda for several years tried to move both share and margin, but since the Walmart acquisition the focus has been on greater sales and gaining market share.

New managerial techniques continue to emerge and be applied in the food distribution sector, particularly in retailers. During the 1990s, category management became widely accepted as the managerial solution to integrating marketing and trading (Ajula & Boitoult, 2002; IGD, 2002c). This approach enabled the creation of multifunctional teams that focused on a specific category and were responsible for the buying, marketing and merchandising of the category; the team often included representatives from suppliers. A different type of integration has been sought through the idea of efficient consumer response (ECR) (Boitoult *et al.*, 1998). In this case, an attempt is made to co-ordinate the processes across the various firms in the chain in order to focus on increasing the effectiveness of activities

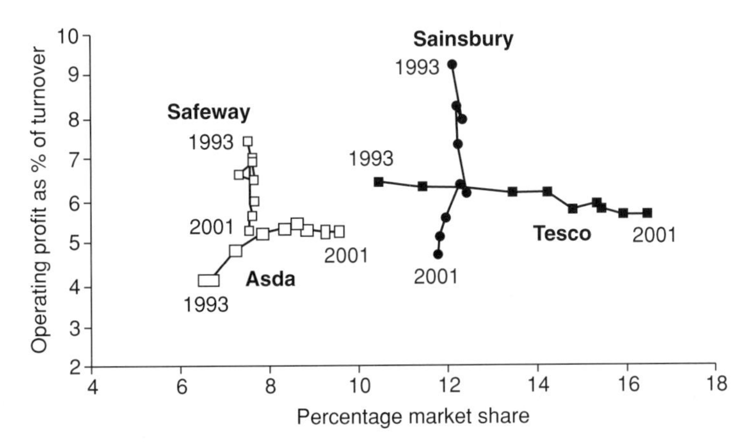

Figure 8.6 Profit margin versus market share (1993–2001) for four leading retail firms (Source: company data).

at store level, in terms of efficiency in ranges, efficient promotions, efficiency in replenishment, and efficient new product introductions.

This channel perspective has resulted in a number of specific managerial concepts being implemented; those likely to be particularly important by 2005 are sales based ordering and factory gate pricing. With sales based ordering there is a steady flow of information from retailer to supplier on what is being sold. This replaces the batching of an order and its delivery, for example, every week. The continuous information flow enables the supplier to optimise production schedules and also means that stores receive replacement items via multiple deliveries per day, replacing the peaked nature of previous replenishment practices. Items move more quickly through the total cycle of ordering, production and delivery. In factory gate pricing the retailer takes the responsibility for the transport of the item from factory to distribution centre and store. Certainly for larger retailers, this enables them to optimise transport schedules and reduce transport costs. The managerial response to advances in technology enables new efficiencies to be achieved. Developments in radio frequency activated tags on cases and trays used to deliver products to stores, electronic shelf-edge labels that allow prices to be changed instantaneously and information about the item to be displayed, and customer self-scanning equipment, are technologies that are likely to become more widespread. In considering the future shape of the food distribution sectors it is necessary to consider these managerial innovations and how they will have impacts on retailing, wholesaling and catering.

Finally in looking to the future, it is important to realise that food distribution is not immune from the processes of internationalisation (Wrigley & Lowe, 2002). These processes affect the sourcing of items and also the operation of outlets. Some catering companies, e.g. McDonalds and Pizza Hut, have a substantial history of having sourced and operated internationally. These internationally orientated firms were the exceptions, but they are rapidly becoming the norm. For the future we can forecast an increase in the number of overseas buying offices, particularly in Asia, being operated by food distributors. We can expect an increase in the international flow of food products into the UK, with a consequential rise in importance of international transport by food distribution firms (Downing, 2002). We can also anticipate more foreign owned firms entering the UK food distribution market. In retailing, there is already a major presence by Walmart, Aldi, Lidl and Netto; in wholesaling there is Metro who operate Makro; in catering there is an extensive network by fast-food firms which is increasing with investments such as that of McDonalds in Pret à Manger.

Food distribution in the UK is a highly diverse part of the food chain. Although there are a small number of large and increasingly dominant firms, levels of competition remain high because consumers change their needs quite rapidly, so providing opportunities for small firms to emerge. Food distribution is increasingly international. The large international firms are bringing new managerial concepts to bear on increasing the level of productivity whilst still providing a high level of service to a spatially very disaggregated market. In consequence, the retail, wholesale and catering sectors today bear little resemblance to what was in place 40 years ago, and with the increasing speed of change the sectors will be transformed again within the next 20 years.

Study questions

1. What do you consider to be the major differences in the markets served by retailers, wholesalers and caterers?
2. What are the benefits of operating through multiple formats?
3. Why has branding become so important for food distribution firms?
4. What scale-related benefits accrue to food distribution firms? Are there resulting customer benefits or costs?
5. Is technology change in food distribution a source of sustainable competitive advantage?

References

Ajula, E. & Boitoult, L. (2002) *Category Management: Which Way Now?* Institute of Grocery Distribution, Watford.

Akehurst, G. (1983) Concentration in retail distribution: measurement and significance. *Service Industries Journal* **3**, 161–179.

Boitoult, L., McGrath, M., Offer, C., Sheldon, D., Short, A. & Wakeling, B. (1998) *ECR: An Industry Response*. Institute of Grocery Distribution, Watford.

Burt, S. (2000) The strategic role of retail brands in British grocery retailing. *European Journal of Marketing* **34**(8), 875–900.

Burt, S. & Sparks, L. (1994) Structural change in grocery retailing in Great Britain: a discount orientation? *International Review of Retail Distribution and Consumer Research* **4**, 195–217.

Burt, S. & Sparks, L. (1997) Performance in food retailing: a cross national consideration and comparisons of retail margins. *British Journal of Management* **8**, 133–150.

Clarke, I. (2000) Retail power, competition and local consumer choice in the UK grocery sector. *European Journal of Marketing* **34**, 975–1002.
Clarke, R., Davies, S., Dobson, P. & Waterson, M. (2002) *Buyer Power and Competition in European Food Retailing*. Edward Elgar, Cheltenham.
Coca Cola (2001) *The Store of the Future*. Coca Cola Research Group Europe, London.
Competition Commission (2000) *Supermarkets: A Report on the Supply of Groceries from Multiple Stores in the United Kingdom*. The Stationery Office, London.
Coopers & Lybrand (1996) *The Future for the Food Store*. Coca-Cola Research Group Europe, London.
Davies, K., Gilligan, C. & Sutton, C. (1985) Structural changes in grocery retailing: the implications for competition. *International Journal of Physical Distribution and Materials Management* **115**, 3–48.
Dawson, J.A. (1995) Food retailing and the food consumer. In: *Food Choice* (Marshall, D.W., ed.), pp.77–104. Blackie, London.
Dawson, J.A. & Shaw, S.A. (1990) The changing character of retailer–supplier relations. In: *Retail Distribution Management* (Fernie, J., ed.), pp. 19–39. Kogan Page, London.
DETR (1998) *The Impact of Large Food Stores on Market Towns and District Centres*. The Stationery Office, London.
Dobson, P., Waterson, M. & Chu, A. (1998) *The Welfare Consequences of the Exercise of Buyer Power*. Research Paper 16. Office of Fair Trading, London.
Downing, N. (2002) *The Future of Global Sourcing*. Institute of Grocery Distribution, Watford.
Guy, C. (1996) Corporate strategies in food retailing and their local impacts: a case study of Cardiff. *Environment and Planning* **A28**, 1575–1602.
Hunt, I.J. (1983) Developments in food distribution. In: *The Food Industry: Economics and Policies* (Burns, J., McInerney, J. & Swinbank, A. eds), pp. 127–141. Heinemann, London.
IGD (2002a) *Grocery Retailing 2002*. Institute of Grocery Distribution, Watford.
IGD (2002b) *Grocery Wholesaling 2002*. Institute of Grocery Distribution, Watford.
IGD (2002c) *The Future of UK Grocery Retailing*. Institute of Grocery Distribution, Watford.
Laaksonen, H. & Reynolds, J. (1994) Own branding in food retailing across Europe. *Journal of Brand Management* **2**(1), 37–46.
Moir, C. (1990) Competition in the UK grocery trades. In: *Competition and Markets: Essays in Honour of Margaret Hall* (Moir, C. & Dawson, J.A.), pp. 91–118. Macmillan, London.
Ogbonna, E. & Wilkinson, B. (1998) Power relations in the UK grocery supply chain. Developments in the 1990s. *Journal of Retailing and Consumer Services* **5**, 77–86.
ONS (2002) *Annual Business Inquiry*, Office of National Statistics, Newport (www.statistics.gov.uk/abi).
Seth, A. & Randall, G. (1999) *The Grocers*, Kogan Page, London.
Sparks, L. (1993) The rise and fall of mass marketing? Food retailing in Great Britain since 1960. In: *The Rise and Fall of Mass Marketing* (Tedlow, T.L.S. & Jones, G. eds), pp. 58–92. Routledge, London.
Wileman, A. & Jary, M. (1997) *Retail Power Plays: From Trading to Brand Leadership*. Macmillan Business, Basingstoke.
Wrigley, N. (1992) Sunk capital, the property crisis and the restructuring of British food retailing. *Environment and Planning* **A24**, 1521–1527.
Wrigley, N. (1994) After the store wars. Towards a new era of retail competition. *Journal of Retail and Consumer Services* **1**, 5–20.
Wrigley, N. (2001) Local spatial monopoly and competition regulation: reflections on recent UK and US rulings. *Environment and Planning* **A33**, 189–194.
Wrigley, N. & Lowe, M. (2002) *Reading Retail*. Arnold, London.

Chapter 9

Partnerships and Alliances in UK Supermarket Supply Networks

R. Duffy & A. Fearne

Objectives

(1) To explain the rationale for greater collaboration in supermarket supply chains.
(2) To discuss the main areas in which collaboration takes place.
(3) To present some insights from the literature on buyer–supplier relationships and the impact of supply chain collaboration on the performance of UK supermarkets and their suppliers.

Introduction

The 1990s was the decade during which strategic management researchers woke up to the paradigm shift that was emerging in the way firms sought to compete. The realisation that sustainable competitive advantage could not be achieved by firms in isolation, but through the engineering and management of efficient and effective supply chains, in collaboration with their customers and suppliers, opened the way for numerous researchers, from strategic, operations, logistics, manufacturing and marketing management, as well as purchasing and finance, to press the case for 'co-opetition' (Brandenburger & Nalebuff, 1996) rather than competition as the way forward (see also Houlihan, 1988; Bhote, 1989; Carlisle & Parker, 1989; Lewis & Thomas, 1990; Christopher, 1992; Sako, 1992; Harrison, 1993; Kay, 1993; Lamming, 1993; Hines, 1994; Saunders, 1994; Gattorna & Walters, 1996; Christopher, 1997; Lambert *et al.*, 1998).

The basic argument of this chapter is that business success will be derived from companies managing and enhancing the total performance of the supply chain, for the purpose of delivering improved value to customers. Waste is normally seen as the major enemy; closer and long-term working relationships, even partnerships, with suppliers and customers at all levels in the chain are recommended in order to deliver exceptional value to customers at minimal cost and as fast as possible.

Thus, companies are seeking to construct ever more efficient and responsive supply chains as the nature of competition moves from firm against firm to supply chain against supply chain.

The food industry has been slow to embrace the partnership philosophy and it is only in recent years that supply chain management has made its way onto the boardroom agendas of the world's leading food manufacturers (Fearne *et al.*, 2001). Progress has been particularly slow upstream, where a distinct lack of trust between trading partners has made the task more difficult and the process longer. However, the launch of the efficient consumer response (ECR) initiative, initially in the US grocery industry and later throughout Europe, provided the catalyst for the paradigm shift that we now see in the management of the food supply chain, with adversarial trading relationships slowly being replaced by co-operation and co-ordination, facilitated by a willingness to exchange information of both strategic and operational importance. As a result, the world's leading food manufacturers are reducing lead times, inventory levels and new product development (NPD) cycle times, delivering a more effectively managed range of carefully targeted products and services to increasingly diverse groups of consumers, at substantially lower costs.

ECR is based on the premise that many business practices and attitudes within the food industry are counterproductive, with firms seeking to maximise their own efficiency and profitability by passing problems and costs up or down the supply chain to their trading partners. Therefore, the fundamental aim of ECR is to apply a total systems view and encourage firms to work together to remove unnecessary costs from the supply chain and to add value to products by identifying and responding to consumer needs more effectively. Because ECR relies on a seamless flow of information throughout the supply chain, its benefits are dependent on a move away from traditional confrontational relationships to relationships based on co-operation and trust (Wood, 1993; IGD, 1996; Lamey, 1996; Fiddis, 1997; Mitchell, 1997).

Whilst most initial ECR activity was carried out between the multiple retailers and large branded manufacturers, the concepts were soon extended to commodity sectors, such as fresh fruit and vegetables and fresh meat. These product categories are almost exclusively own-label and have become what retailers describe as 'destination categories', because they are product categories for which shoppers will switch stores (Corstjens & Corstjens, 1995; Fearne & Hughes, 1999). This is because shoppers accord great importance to the quality, range, availability and price of fresh food items when selecting their store of first choice (Hughes & Merton, 1996).

The establishment of strong own-label products requires retailers to gain increased control over the supply chain to ensure that product quality and availability reflect the desired positioning of the retail brand. Therefore, it is in the interests of the retailers to develop closer relationships with the suppliers of these products, who are increasingly taking responsibility for sourcing year round supplies and ensuring compliance with food safety and quality assurance specifications. The emergence of retailer-driven quality assurance schemes, from farm to fork, initially in the UK but increasingly throughout the developed world, is testimony to the growing interest in supply chain collaboration. Indeed, the labels given to many of these schemes (e.g. Sainsbury's Partnership in Livestock, Tesco's Beef Club and Marks and Spencer's Select Beef) suggest that supply chain collaboration is now established as the preferred mode of conducting business in the retail food chain.

Structure of UK supermarket supply chains

The UK food retailing industry is dominated by nine major multiple retailers: Tesco, Sainsbury, Asda, Safeway, Somerfield, Marks and Spencer, Morrisons, Waitrose and Iceland. Of these nine major retailers, the top four, Tesco, Sainsbury, Asda and Safeway, account for over 62% of the total grocery market (TNS, 2001). The growth of the UK multiple retailers means that the purchasing power of the food retailing industry is concentrated in the hands of a relatively small number of retail buyers.

UK multiple retailers have rationalised their supply base dramatically in recent years as they have recognised that they can reduce purchasing costs by dealing with fewer suppliers. Supplier rationalisation has meant that the major supermarkets now deal with just a handful of suppliers in each product area. These suppliers are typically large pre-packers or processors that have geared up to meet the needs of the multiples. These suppliers are the key link between farmers and the supermarkets, and increasingly take responsibility for sourcing supplies, liasing with farmers and building global networks for year round supplies. In return these suppliers have been rewarded with volume growth.

The growing importance of the retailer–supplier link in the supply chain has resulted in closer relationships between these two parties. Key drivers for the development of closer relationships have included the need for retailers to reduce risk in response to the requirements of the Food Safety Act (1990), the need to develop quality own-label products in an effort to differentiate themselves from their competitors, and the need to reduce costs to remain competitive in a highly concentrated market. However, the most notable driver in recent years has been the introduction of the industry initiative efficient consumer response (ECR).

The premise of ECR is that by working together costs can be removed from the supply chain and value can be added to products by identifying and responding to consumer needs more effectively. In trying to respond more effectively to consumer demand a number of retailers have introduced the concept of category management to their suppliers. This requires suppliers to have a greater understanding of the final consumer and the market situation so that they supply retailers with products that the consumer wants, and in this way sales and profits in the category can be maximised. In some cases, retailers have nominated one supplier to be the category leader or 'category captain'. This supplier might be the sole supplier in a category to a retailer or might be the main link between the retailer and other suppliers. The importance of the retailer–supplier link means that supermarkets have little or no direct contact with their grower or farmer base. Increasingly,

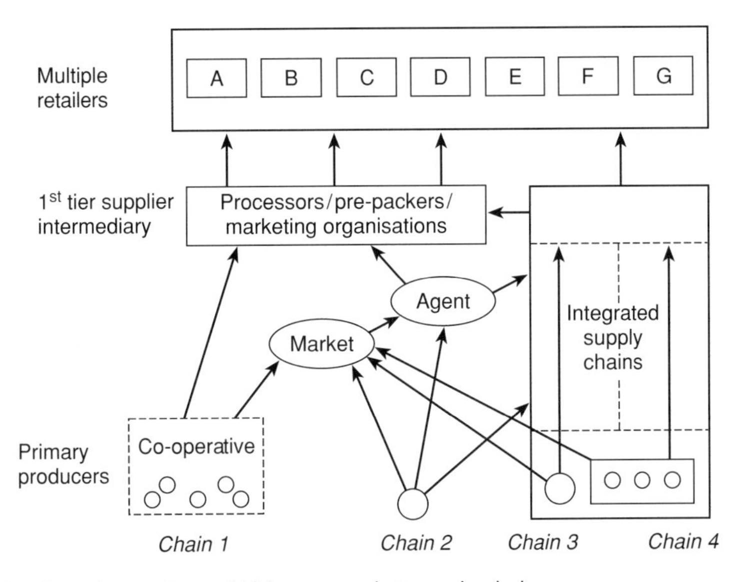

Figure 9.1 Generic structure of UK supermarket supply chains.

retailers rely on these first tier suppliers to co-ordinate the supply obtained from primary producers and ensure that these growers and farmers comply with the specifications set by the retailers concerning food safety and quality assurance.

A generic supermarket supply chain structure is presented in Figure 9.1, which illustrates the diversity of channels upstream from the first tier suppliers (processors, packers, marketing organisations). Supply chain 1 comprises growers or farmers that are members of a co-operative or producer group. These primary producers will sell their output to a processor, pre-packer or marketing agent who then supplies the multiple retailers. Some of the primary producers' output will also be sold to the wholesale market. Agents or brokers might purchase this output from the wholesale market, particularly in the meat sectors, and supply the first tier suppliers who supply the multiple retail market. This kind of supply chain is common in the meat and fresh produce sectors, where wholesale markets remain but are diminishing in importance and where producer controlled organisations have developed to benefit from economies of scale, without taking the significant step into value-added processing. Top fruit, soft fruit, beef and lamb would be the most prevalent sectors in which these supply chains continue to represent a significant proportion of supermarket supplies.

Supply chain 2 comprises primary producers that might be viewed as being 'outside' the supermarket supply chain. This is because they sell their produce through the wholesale market or to agents that do not supply the multiple retailers directly. This kind of supply chain represents an extremely small proportion of retail sales, because retailers have pursued a policy of supply base rationalisation, preferring to deal with producer groups and large scale units, whilst simultaneously reducing their dependence on wholesale markets due to problems associated with quality and traceability. However, there remains a significant number of small scale livestock, dairy and fruit farmers who find themselves in this supply chain, for whom the future looks particularly bleak, notwithstanding the current interest in farmers' markets.

Supply chain 3 comprises large primary producers that pack and market their own product and supply the multiple retailers directly. However, it is likely that for some of the larger multiple retailers the large independent farmers will supply these customers indirectly through a first tier supplier. These farmers will also send some of their output to the wholesale market and might supplement their own supply with output purchased from other farmers. This kind of supply chain is almost exclusively found in the fresh produce sector, in which large scale producers of brassicas, field vegetables and salad crops have been encouraged to integrate vertically and focus increasingly on supplying specific (i.e. differentiated) products on an exclusive basis to supermarkets seeking a unique selling proposition in these categories.

Finally, supply chain 4 comprises vertically integrated supply chains in which companies are involved in the production, processing, packing and marketing of their output. In some cases, these companies may draw the balance of their supply from independent farmers or from brokers. These companies supply the multiple retailers directly or might market some of their output through another first tier supplier.

These types of supply chains are prevalent in the meat sector, and to a lesser extent in the dairy sector, where the options for differentiation and value-adding activities are greater than in the fresh produce sector, as is the level of investment required.

The nature of trading relationships in these supply chains and the degree of collaboration between trading partners vary considerably between sectors and over time. Generally speaking, the more differentiated the product and/or service, the greater the collaboration, particularly for own label products where supply chain 4 has emerged as the dominant structure in which dedicated suppliers work increasingly closely and exclusively with individual retailers. The temporal perspective is important because trading relationships evolve over time and are largely determined by strategic objectives. Thus, as UK supermarkets have attempted to develop their own label offerings to compete head on with branded substitutes and attract new shoppers into destination categories, so they have recognised the need to invest in strategic supply networks in support of their corporate objectives.

The nature of collaboration in supermarket supply chains

Collaboration between firms in the food industry may be either vertical (upstream or downstream) or horizontal in nature. Gattorna & Walters (1996) state that although the terms 'strategic partnerships' and 'strategic alliances' have been used interchangeably, partnerships refer to vertical relationships between firms in different stages of the supply chain (i.e. retailer and manufacturer), whereas alliances refer to horizontal relationships between firms at the same stage of the supply chain (i.e. farmer co-operatives and producer groups). The focus here is on the former, as they are representative of the paradigm shift in the nature of competitive strategy described previously.

While the concept of supply chain collaboration has become popular, Tan *et al.* (1999) state that, in practice, few examples of truly integrated supply chains exist because it is often not feasible or appropriate for a firm to develop close linkages with all firms throughout the length of the supply chain (Lambert *et al.*, 1998). Instead, as the supply chain consists of a series of vertical inter-organisational relationships of a dyadic nature, a common approach to supply chain management is dyadic management (Harland, 1995). Firms using this approach focus only on developing close linkages with those channel members with whom they have immediate contact; this spreads throughout the supply chain so that it becomes progressively integrated and managed, dyad by dyad. Such an approach is popular because it does not require total central co-ordination and control which can be difficult and costly to implement. Therefore supply chain collaboration involves identifying the supply chain members with whom it is critical to link as a firm is unlikely to have the resources to develop close linkages with all other firms in its supply chain (Lambert *et al.*, 1998). This is indeed the way in which supply chain partnerships have evolved in the retail food sector, with retailers focusing on their relationship with suppliers, be they manufacturers of branded or own label grocery products or processors/packers of commodity lines, and the latter developing their relationship(s) with raw material and ingredient suppliers.

Vertical supply chain partnerships, such as those developed between food retailers and their suppliers, are not necessarily exclusive and rarely contractual relationships. Instead they may be described as '. . . some arrangement between buyer(s) and seller(s), entered into freely, to facilitate a mutually satisfying exchange over time, and which leaves the operation and control of the businesses substantially independent' (Hughes, 1994). There are four key aspects of this definition:

(1) Partnerships are entered into 'freely'. Partners do have a choice, although the upstream options may be becoming increasingly limited, given the dominance of supermarkets in the retail food chain.
(2) Partnerships must offer 'mutual' benefits. These are many and varied and their distribution is one of the key problem areas in the life cycle of supply chain partnerships.
(3) These benefits occur 'over time' and what distinguishes partnerships from purely transactional relationships, such as open market 'spot trading', is the time dimension of the payback, which we generally associate with investment and a strategic, rather than opportunistic, approach to the way we manage the relationships with customers and suppliers.
(4) Partners remain 'substantially independent' and what distinguishes vertical partnerships from vertical integration is the lack of equity sharing and the absence of contractual obligations, which allows much greater flexibility in the nature and scope of the relationship between partners.

Thus, supply chain partnerships represent efforts to achieve goals that individual firms acting alone could not attain easily. For example, in the context of the retail food chain, initiatives such as ECR seek to achieve cost savings through reduced inventory and waste, and increased revenue through more effective promotions, product assortments and new product introductions. Clearly, this is not possible unless retailers share information on consumer demand and plan promotions in collaboration with suppliers to ensure availability of stock without recourse to markdowns. The advent of collaborative planning, forecasting and replenishment (CPFR) systems, through which retailers and manufacturers share real-time information on sales and stocks by store and regional distribution centre (RDC), is a major breakthrough in this regard, with retailers and manufacturers citing significant benefits in customer service, inventory management and waste reduction (Stank *et al.*, 1999a,b).

Methods of co-ordinating the individual links in the food supply chain can be conceptualised as existing on a continuum. The conventional economic view is that, at one end of the continuum, exchange between two firms is conducted by open market mechanisms such as auctions where interaction between the buyer and seller is limited to the individual transaction. At the other end of the continuum, co-ordination in the supply chain is achieved through vertical integration whereby successive links in the supply chain are owned by the same firm. Between these two extremes lie a variety of structures in which the market mechanism is modified through some kind of formal or informal contractual arrangements between the parties involved (Williamson, 1975). These extreme forms of co-ordination have now decreased in importance and instead it is more common for exchange to occur within the framework of long-term buyer–supplier relationships.

Theory of buyer–supplier relationships

Buyer–supplier relationships range from primarily transactional ties, where the focus is limited to the timely exchange of products for highly competitive prices, to collaborative relationships or partnerships where firms form strong and extensive ties, with the intent of lowering total costs and/or increasing value, thereby achieving mutual benefit (Jackson, 1985; Carlisle & Parker, 1989; Webster, 1992; Kanter, 1994).

Spekman *et al.* (1998) state that co-operation, whereby firms exchange bits of information and engage in longer term contracts, has now become the threshold of interaction for firms. The next level of intensity is co-ordination, whereby both specified workflow and information are exchanged in a manner that makes seamless the traditional communication linkages between trading partners. In most cases they state that firms have already achieved co-operation and co-ordination with key segments of their suppliers and customers; however, they add that trading partners can co-operate and co-ordinate certain activities but still not behave as true partners and collaborate. This is because collaboration requires high levels of trust, commitment, information sharing and a common vision of the future. Therefore firms in collaborative partnerships engage in joint planning and processes

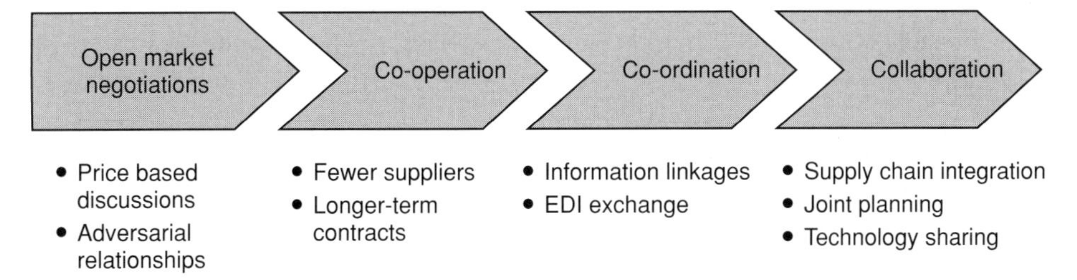

Figure 9.2 Transition from open market negotiations to collaboration (Source: Spekman *et al.*, 1998).

Box 9.1 Key extremes of buyer–supplier relationships.

Traditional arm's-length relationships
Short-term focus on individual transactions
Buying decision made on price
Many suppliers
Low interdependence
Haphazard production and supply scheduling
Limited communication restricted between sales and purchasing
Little co-ordination of work processes
Relationship specific investments avoided
Information is proprietary
Clear delineation of business boundaries
Use of threats to resolve disputes
Unilateral improvement initiatives
Separate activities
Dictation of terms by more powerful firm
Adversarial attitudes/combat
Conflicting goals
Behave opportunistically
Act only in own interest
Win–lose orientation

Supply chain partnerships
Commitment to long-term relationships
Buying decision made on value
Fewer selected suppliers
High interdependence
Order driven production and supply
Open communication facilitated by multi-level/multifunctional relationships
Integration/co-ordination of work processes
Increases in relationship specific investments
Information is shared
Creation of inter-company teams
Joint problem solving approach to conflicts
Continuous joint improvement sought
Engage in joint activities
Joint decision making
Co-operative attitudes/teamwork
Compatible goals
Mutual trust exists
Act for mutual benefit
Win–win orientation

beyond levels reached in less-intense trading partnerships (Figure 9.2). Numerous researchers have suggested characteristics that distinguish between these two extremes (Landeros & Monczka, 1989; Heide & John, 1990; Burdett, 1992; Imrie & Morris, 1992; Stuart, 1993; Matthyssens & Van den Bulte, 1994; Kanter, 1994; Ellram & Hendrick, 1995; Joseph *et al.*, 1995; Schroder & Mavondo, 1995; Mudambi & Schrunder, 1996; Spekman *et al.*, 1998). These key characteristics are summarised in Box 9.1.

A review of the theory of inter-organisational relationships in the marketing channel literature (Duffy & Fearne, 2002) identifies three dimensions of buyer–supplier relationships that are key influences on the performance thereof.

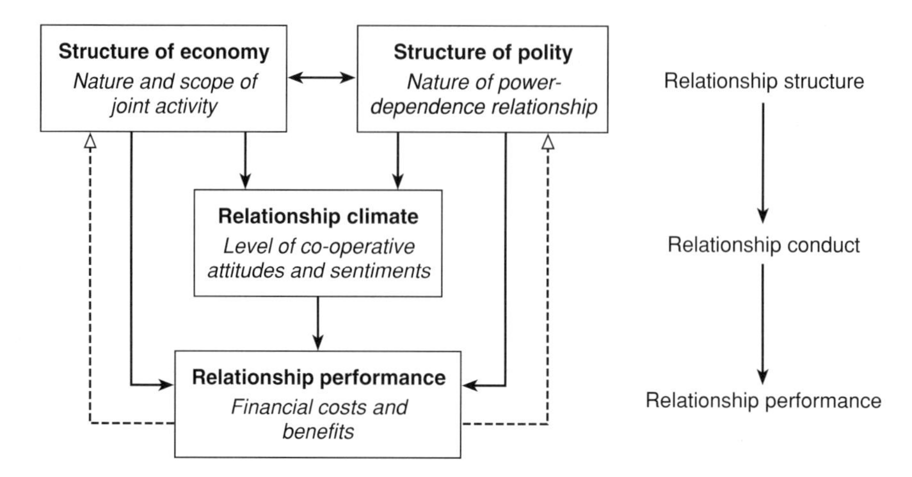

Figure 9.3 Theoretical framework of buyer–supplier relationships.

(1) The structure of the relationship economy that is the nature and scope of joint activities undertaken.
(2) The structure of the relationship polity that is the nature of the power dependence relationship.
(3) The relational norms in terms of the dominant attitudes and sentiments that exist.

These three factors combine to positively influence performance of dyadic relationships between strategically important buyers and suppliers, those who engage in joint activities, develop mutual dependence over time and exhibit trust and functional methods of conflict resolution (Figure 9.3).

Nature and scope of joint activity

A key dimension that should receive attention when managing supply chain relationships is the tangible physical and technical components of the relationship (Lambert *et al.*, 1998). These tangible components can be defined in terms of the task related flows of activities, resources and information that are used to support and co-ordinate activities in the trading relationship.

In UK supermarket supply chains, the ECR initiative has encouraged trading partners to work together and seek improvements in three primary areas. These are: (1) understanding and responding to consumer demand (known as demand or category management); (2) operating efficiently through the supply chain (efficient replenishment or supply chain management), and (3) collaborative working, through the supply chain by fully utilising enabling technologies and practices (IGD, 1996; 1999). Applied together these three areas offer practitioners a supply chain that works 'better, faster and cheaper' plus a more effective focus on ever shifting consumer demands, leading to incremental sales and profit growth (IGD, 1996).

Category management addresses a process for demand creation and is concerned with the strategic management of product groups through trade partnerships, aiming to maximise sales and profits by satisfying consumer needs. In terms of practical implementation, category management focuses on the three improvement concepts of efficient assortment, efficient promotion and efficient new product introductions. Efficient assortment is concerned with maximising profitability through the effective use of store space and optimising the product mix to match consumer needs. Through fully understanding consumer buying behaviour, unproductive and duplicate lines can be eliminated from the shelves, whilst still offering sufficient consumer choice. In addition, product ranges need to be designed for different store sizes, regional preferences and catchment areas.

Consumer and trade promotions are used to stimulate purchase and create excitement in store. However, ECR recognises that there is huge potential to make promotions more efficient as these often have negative effects on costs throughout the supply chain. In particular, promotions are often not well planned, so it is not clear whether the aim is to encourage trial, grow market share, meet sales targets, boost short-term profits, enhance a value-for-money image, or respond to competitive activity. If the aims are not clear, it is difficult to assess whether the promotion has been successful. An evaluation of promotional activity is important because of the extra costs incurred, such as in designing and making special packaging. In addition, promotions often encourage consumers to stockpile rather than consume more, so many sales boosts are short-term and do not translate into increased profits.

Category management also focuses attention on new product introductions, which are important to maintain consumer interest in the category and to drive new sales. The area of product introductions offers the potential for improvement because over 80% of new product launches fail to sustain a market presence after two years (IGD, 1997).

A key goal of ECR is to encourage retailers and suppliers to work together to discover which new product idea will be successful before the expense of the launch. This can be done by ensuring that the new product idea adds real value to the category, through meeting an identified need, attracting new consumers and helping to increase category performance.

The second key area of activity in which firms can work together for improvement is management of the logistics function, which covers activities such as store replenishment, preparation of purchase orders, inventory control, warehousing and transport flows. Isolated attempts to maximise supply chain efficiency are often undermined or contradicted by the actions of trading partners. Therefore companies need to see their operations from the perspective of the virtual enterprise, where information about product flows is transparent and there are no barriers to the flow of goods created by organisational boundaries. This means that information about sales flows naturally from retailers to suppliers. Traditionally each party has guarded this information so that pure market signals created by actual consumer demand have been distorted.

Efficient replenishment has been the traditional starting point of ECR activity between retailers and manufacturers as it offers the quickest and most obvious benefits. For example, efficient replenishment offers companies the opportunity to reduce costs by lowering inventory levels, which frees up cash, reduces storage requirements and reduces handling and warehouse operations. In addition, improved information about product flows enables firms to make better use of delivery vehicles and achieve better demand forecasting, which improves on-shelf product availability and reduces product waste (i.e. out-of-date products). The introduction of 'just-in-time' deliveries for store replenishment also enables retailers to reallocate space from back-stage storage operations to sales footage. Such space reallocation is valuable to retailers because planning restrictions for new stores have become more restrictive.

The third area of joint activity is the use of information sharing technologies and on-line computer linkages, because the availability of relevant, timely and usable information between trading partners is critical to the success of any initiative that seeks to achieve collaboration. Firms who are operating in a collaborative manner will also engage in open communication where each partner can reveal its own needs, capabilities and limitations to the other (Spekman & Salmond, 1992). Firms engaging in collaborative relationships are also likely to have direct contact with their operational counterpart, which helps to ensure that problems can be sorted out quickly and easily.

Nature of the power dependence relationship

In a supply chain context, dependence is defined as the extent to which a trade partner provides important and critical resources for which there are few alternatives (Cadotte & Stern, 1979; Gundlach & Cadotte, 1994). Therefore, to the extent that valued resources and outcomes are available outside the relationship, dependency is minimised and partners are more easily pulled away from their established relationship.

The notion of the power dependence relationship is illustrated in Emerson's (1962) theory of dependence, which states that power resides implicitly in the other's dependency. The premise of Emerson's theory is that the power of firm A over firm B is equal to and based upon the dependence of firm B upon firm A. In exchange relationships, both participants are to some degree dependent on each other and so they are said to be interdependent. The level of each partner's dependence, or the structure of reciprocal dependence, characterises the nature of interdependence in the relationship.

It is important to understand the structure of interdependence, as the degree of partnership that can develop between two firms is determined by the relative power and interdependence of the parties involved (Kearney, 1994; Gattorna & Walters, 1996); for example, if both firms place a high value on the relationship they will be willing to work to develop and improve it for their mutual benefit. Therefore the higher the level of interdependence, the more appropriate is a strategic partnership, as collaboration requires both parties to acknowledge that the resultant gain cannot be easily replicated in another trading relationship (Gattorna & Walters, 1996).

The structure of interdependence also has implications for the performance outcomes achieved in the relationship and the distribution of the costs and benefits of the exchange. With regard to the nature of asymmetry in the relationship, two points of view exist regarding the relationship between dependence and performance. The opportunistic perspective suggests that the possession of more power (i.e. less dependence) will manifest exploitative tendencies and encourage action to gain a disproportionate share of resources from a less powerful partner (Beier & Stern, 1969; Buchanan, 1986; Noordewier *et al.*, 1990; Gundlach & Cadotte, 1994). On the other hand, the benevolent perspective suggests that those with the greatest power are able to manipulate other members to act in ways that achieve greater positive results for the whole system and encourages channel member co-operation (Beier & Stern, 1969). However, the exercise of control will be tolerated only if the benefits of such control are understood, realised and shared equitably by the channel member involved (El-Ansary, 1979).

Concentration in the food retail sector means that retailers are in a position of power relative to their suppliers. However, the structure of interdependence in retailer–supplier relationships varies according to the market structure and the type of product supplied because this determines available choice and the power of prospective partners in a market. For example, branded food manufacturers dominate certain product categories and so tend to have more balanced power relationships with retailers than suppliers of commodity products. If retailers do not stock these leading brands they risk losing customers to retailers that do, which gives branded manufacturers a source of countervailing power. However, suppliers of commodity products are less able to differentiate their products at the consumer level and so are in a weaker bargaining position as they can only differentiate themselves in terms of price. Although the effect of continued supplier rationalisation in all product sectors means that retailers are increasingly dependent upon a few key suppliers who have geared up to meet their needs, retailers are still able to switch volumes between suppliers. Hence, suppliers of commodity items are often forced to accept low prices in order to get volume growth, which does little to improve their immediate and long-term financial performance.

Suppliers in commodity sectors therefore need to find ways to add value to the product or service, so that they can differentiate themselves from their competitors, and appropriate value for themselves rather than passing all value on to the retailer. Value creation is achieved through innovation, and while a lack of product innovation is a feature of commodity markets, consumers' demand for greater convenience offers significant opportunities for growth in certain areas, such as ready prepared vegetables (Fearne & Hughes, 1999). However, innovation is difficult to achieve and exploit in commodity sectors which offer low margins for suppliers and in which the rewards for first movers on new products (varieties, preparation and packaging) are limited and short lived. Therefore suppliers also need to differentiate themselves from their competitors in terms of the service they offer to retailers. One area suggested by Fearne and Hughes (1999) is product knowledge, as this is one of the few areas where suppliers can and should have an advantage over their customers. Thus, market knowledge offers suppliers one of the few remaining sources of countervailing power in commodity sectors.

Level of co-operative attitudes and sentiments

Conflict and co-operation are the two dominant sentiments that regulate exchange relationships (Stern & Reve, 1980; Skinner *et al.*, 1992). The type of sentiments that exist in relationships can influence the development of partnerships, because to be able to work together effectively, firms must be willing to share information. Partnerships also require considerable investments of time, resources and effort. Therefore collaborative business approaches require a movement away from traditional confrontational relationships to those based on co-operation, openness, trust and commitment.

Researchers such as Morgan & Hunt (1994) and Mohr & Spekman (1994) state that the presence of commitment and trust is central to successful relationships, as their presence leads directly to co-operative behaviours directed towards collective as opposed to individual goals. Commitment refers to the willingness of trading partners to exert effort on behalf on the relationship and reflects the firm's intent to remain in the relationship; without commitment, firms will not be willing to make investments of time or capital in the relationship, which will hinder its development and effectiveness.

Inter-organisational trust also affects the development of partnerships as it encompasses two elements that are essential if two firms are to work together in collaborative partnerships. The first, trust in the partner's honesty, is the belief that the partner stands by its word, fulfils promised role obligations and is sincere, while the second, trust in the partner's benevolence, is the belief that the partner is interested in the firm's welfare and will not take unexpected actions that will negatively affect the firm (Kumar *et al.*, 1995). Firms that do not trust each other will be discouraged from sharing information and making investments in their relationships. Therefore a lack of trust hinders the development of partnerships and has been cited by Sherman (1992) as the biggest stumbling block to the success of alliances.

The historical lack of trust at any level of the food chain has undoubtedly been a major barrier to the adoption of supply chain management within supermarket supply chains. Indeed, in recent years, UK supermarkets have been subjected to intense scrutiny with respect to the dominant sentiments and attitudes adopted by their buyers when dealing with suppliers. Most notably, in October 2000, the Competition Commission published the results of a high profile enquiry into supermarket behaviour, in which numerous abuses of power were noted. These included requests from some supermarket suppliers for various non-cost related payments or discounts; imposing charges; making changes to contractual arrangements without adequate notice; and unreasonably transferring risks from the supermarket to the supplier (Competition Commission, 2000). A significant consequence of the enquiry was the adoption of a voluntary code of practice in March 2002 by the four largest UK supermarkets (Tesco, Sainsbury, Asda and Safeway), designed to improve relationships with their suppliers and establish a relationship climate of trust and mutual commitment and upon which greater collaboration might be established. It is too early to say whether the voluntary code will make a material difference to the dominant sentiments that exist in relationships between retailers and their suppliers, but it clearly signals a recognition that sustainable competitive advantage requires all firms within the supply chain to consider the implications of their own decisions on other supply chain stakeholders, whether upstream suppliers or downstream customers.

Conclusions

Supermarkets have become the dominant players in the UK food retail supply chain in the last 20 years and as a result have drawn considerable attention from the media and the government in terms of the way they use their power in relation to their suppliers. Here, we have attempted to illustrate why the abuse of power is not in the long-term interests of supermarkets and why the majority of them are progressively moving towards more collaborative trading relationships with their suppliers as a source of competitive advantage by reducing supply chain costs and improving customer service and consumer satisfaction through the more effective use of information and the integration of key business processes.

Moving from an adversarial relationship to one based on trust and commitment is never easy and always takes time. Thus, it is our contention that much of the criticism levelled at supermarkets in recent years is misguided and that many of the bad practices that resulted from the abuse of market power in the mid-1980s and early 1990s have been largely competed away, as supermarkets move slowly but steadily to the deployment of their supply chain relationships as a strategic weapon in the quest for sustainable competitive advantage.

Study questions

1. What are the primary motives for UK supermarkets and their suppliers to engage in collaborative activities?
2. In what areas of activity can collaboration between buyers and sellers result in higher levels of efficiency and effectiveness?
3. How does the theoretical literature help us to understand the nature and scope of collaborative arrangements between buyers and sellers?
4. What do you imagine are the greatest barriers to collaboration between supermarkets and their suppliers and how might these be overcome?

References

Beier, F. & Stern, L. (1969) Power in the channel of distribution. In *Distribution Channel: Behavioural Dimensions* (Stern, L. ed.), Houghton Mifflin, Boston, MA.
Bhote, K.R. (1989) *Strategic Supply Management*. Amacom, New York.
Brandenburger, A.M. & Nalebuff, B.J. (1996) *Co-opetiton*. Doubleday, New York.
Buchanan, L. (1986) *The Organisation of Dyadic Relationships in Distribution Channels: Implications for Strategy and Performance*. Stanford University, Stanford, CT.
Burdett, J.O. (1992) A model for customer–supplier alliances, *Logistics Information Management* **5**(1), 25–31.
Cadotte, E.R. & Stern, L.W. (1979) A process model of interorganisational relations in marketing channels. *Research in Marketing* **2**, 127–158.
Carlisle, J. & Parker, B. (1989) *Beyond Negotiation*. Wiley, Chichester.
Christopher, M. (1992) *Logistics and Supply Chain Management*. Pitman, London.
Christopher, M. (1997) *Marketing Logistics*. Butterworth Heinemann, Oxford.
Competition Commission (UK) (2000) *Supermarkets: A Report on the Supply of Groceries from Multiple Stores in the UK*. Competition Commission Report Series, London (http://www.competition/commission.org.uk).
Corstjens, J. & Corstjens, M. (1995) *Store Wars: The Battle for Mindspace and Shelfspace*. Wiley, London.
Duffy, R. & Fearne, A. (2002) *The development and empirical validation of a political economy model of buyer–supplier relationships in the UK food industry*. Discussion Paper 2, November 2002. Centre for Food Chain Research, Imperial College at Wye, Ashford.
El-Ansary, A. (1979) Perspectives on channel system performance. In *Contemporary Issues in Marketing Channels* (Lusch, R.F. & Zinszer, P.H. eds), University of Oklahoma, Oklahoma, Ok.
Ellram, L. & Hendrick, T. (1995) Partnering characteristics: a dyadic perspective. *Journal of Business Logistics* **16**(1), 41–64.
Emerson, R.M. (1962) Power-dependence relations. *American Sociological Review* **27**, 31–41.
Fearne, A. & Hughes, D. (1999) Success factors in the fresh produce supply chain: insights from the UK. *Supply Chain Management* **4**(3), 120–128.
Fearne, A., Hughes, D. & Duffy, R. (2001) Concept of collaboration: supply chain management in a global food chain. In: *Food Supply Chain Management: Issues for the Hospitality and Retail Sector* (Eastham, J., Sharples, L. & Ball, S. eds), Reed Educational and Professional, Oxford.
Fiddis, C. (1997) *Manufacturer–Retailer Relationships in the Food and Drink Industry: Strategies and Tactics in the Battle for Power*. Financial Times Retail and Consumer Publications, London.
Food Safety Act (1990) The Stationery Office, London.
Gattorna, J.L. & Walters, D.W. (1996) *Managing the Supply Chain: a Strategic Perspective*. Macmillan Business, London.
Gundlach, G. & Cadotte, E.R. (1994) Exchange interdependence and interfirm interaction: research in a simulated channel setting. *Journal of Marketing Research* **31**, 516–532.
Harland, C. (1995) The dynamics of customer dissatisfaction in supply chains. *Production, Planning and Control* **6**(3), 209–217.
Harrison, A. (1993) *Just-in-Time Manufacturing in Perspective*, Prentice Hall, London.
Heide, J. & John, G. (1990) Alliances in industrial purchasing: the determinants of joint action in buyer–supplier relationships. *Journal of Marketing Research* **27**, 24–36.
Hines, P. (1994) *Creating World Class Suppliers*. Pitman, London.

Houlihan, J. (1988) Exploiting the industrial supply chain. In *Logistics in Manufacturing* (Mortimer, J. ed.), IFS Publications, London.

Hughes, D. (1994) *Breaking with Tradition: Building Partnerships and Alliances in the European Food Industry*, Wye College Press, Wye.

Hughes, D. & Merton, I. (1996) Partnership in produce: the J. Sainsbury approach to managing the fresh produce supply chain. *Supply Chain Management* **1**(2), 4–6.

IGD (1996) *ECR Process Framework*. Institute of Grocery Distribution, Watford.

IGD (1997) *The Official European ECR Scorecard*. Institute of Grocery Distribution, Watford.

IGD (1999) *Grocery Retailing: the Market Review*. Institute of Grocery Distribution, Watford.

Imrie, R. & Morris, J. (1992) A review of recent changes in buyer–supplier relations. *International Journal of Management Science* **20**(5/6), 641–652.

Jackson, B. (1985) Build customer relationships that last. *Harvard Business Review* **63**, 120–128.

Joseph, W., Gardner, J., Thach, S. & Vernon, F. (1995) How industrial distributors view distributor–supplier partnership arrangements. *Industrial Marketing Management* **24**, 27–36.

Kanter, R.M. (1994) Collaborative advantage: the art of alliances. *Harvard Business Review* **72**(4), 96–108.

Kay, J. (1993) *Foundations of Corporate Success*. Oxford University Press, Oxford.

Kearney, A.T. (1994) *Partnership or Power Play: What is the Reality of Forming Closer Relationships with Suppliers and Customers along the Supply Chain?* Report on the findings of a joint research programme into supply chain integration in the UK. Manchester School of Management, Manchester.

Kumar, N., Scheer, L.K. & Steenkamp, E.M. (1995) The effects of perceived interdependence on dealer attitudes. *Journal of Marketing Research* **32**, 348–356.

Lambert, D., Cooper, M. & Pagh, J. (1998) Supply chain management: implementation issues and research opportunities. *International Journal of Logistics Management* **9**(2), 1–16.

Lamey, J. (1996) *Supply Chain Management: Best Practice and the Impact of New Partnerships*. Financial Times Management Reports, London.

Lamming, R. (1993) *Beyond Partnership: Strategies for Innovation and Lean Supply*. Prentice Hall, London.

Landeros, R. & Monczka, R. (1989) Co-operative buyer–seller relationships and a firm competitive posture. *Journal of Purchasing and Materials Management* **25**(3), 9–17.

Lewis, P. & Thomas, H. (1990) The linkage between strategy, strategic groups and performance in the UK retail grocery industry. *Strategic Management Journal* **11**, 385–397.

Matthyssens, P. & Van den Bulte, C. (1994) Getting closer and nicer: partnerships in the supply chain. *Long Range Planning* **27**(1), 72–83.

Mitchell, A. (1997) *Efficient Consumer Response: a New Paradigm for the European FMCG Sector*. Financial Times Retail and Consumer Publishing, London.

Mohr, J. & Spekman, R. (1994) Characteristics of partnership success: partnership attributes, communication behaviour and conflict resolution techniques. *Strategic Management Journal* **15**, 135–152.

Morgan, R. & Hunt, S. (1994) The commitment–trust theory of relationship marketing. *Journal of Marketing* **58**, 20–38.

Mudambi, R. & Schrunder, C.P. (1996) Progress towards buyer–supplier partnerships: evidence from small and medium-sized manufacturing. *European Journal of Purchasing and Supply Management* **2**(2/3), 119–127.

Noordewier, T.G., John, G. & Nevin, J.R. (1990) Performance outcomes of purchasing arrangements in industrial buyer–vendor relationships. *Journal of Marketing* **54**, 80–93.

Sako, M. (1992) *Prices, Quality and Trust*. Cambridge University Press, Cambridge.

Saunders, M. (1994) *Strategic Purchasing and Supply Chain Management*. Pitman, London.

Schroder, B. & Mavondo, F. (1995) The industrialisation of agriculture: overseas experience and implications for Australia. *Australian Agribusiness Review* **3**, 25–35.

Sherman, S. (1992) *Are Strategic Alliances Working?* Fortune, New York.

Skinner, S.J., Gassenheimer, J.B. & Kelley, S.W. (1992) Co-operation in supplier–dealer relations. *Journal of Retailing* **68**(2), 174–193.

Spekman, R. & Salmond, D. (1992) *A Working Concensus to Collaborate: a Field Study of Manufacturer–Supplier Dyads*. Marketing Science Institute, Cambridge, MA.

Spekman, R., Kamauff, J. & Myhr, N. (1998) An empirical investigation into supply chain management: a perspective on partnerships. *International Journal of Physical Distribution and Logistics Management* **28**(8), 630–650.

Stank, T., Daugherty, P. & Autry, C. (1999a) Collaborative planning: supporting automatic replenishment programs. *Supply Chain Management* **4**(2), 75–85.

Stank, T., Crum, M. & Arango, M. (1999b) Benefits of interfirm co-ordination in food industry supply chains. *Journal of Business Logistics* **20**(2), 21–41.

Stern, L.W. & Reve, T. (1980) Distribution channels as political economies: a framework for comparative analysis. *Journal of Marketing* **44**, 52–64.

Stuart, F. (1993) Supplier partnerships: influencing factors and strategic benefits. *International Journal of Purchasing and Material Management* (fall) 22–28.

Tan, K., Kannan, V., Handfield, R. & Ghosh, S. (1999) Supply chain management: an empirical study of its impact on performance. *International Journal of Operations and Production Management* **19**(10), 1034–1052.

Taylor Nelson Sofres (2001) Taylor Nelson Sofres Data. In *The Grocer Yearbook Supplement 2001*. William Reed Publishing, Sussex.

Webster, F.R. (1992) The changing role of marketing in the corporation. *Journal of Marketing* **56**, 1–17.

Williamson, O. (1975) *Markets and Hierarchies*. Free Press, New York.

Wood, A. (1993) Efficient consumer response. *Logistics Information Management* **6**(4), 38–40.

Chapter 10

New Product Development and Information Technology in Food Supply Chain Management: The Case of Tesco

M. Francis

Objectives

(1) To explain the stages involved in a typical product development process within the food industry.
(2) To describe an innovative application of information technology (IT), and explain why this improved the performance of the above process.

Case background

The fieldwork for the case study that is explained in this chapter was conducted over the two year period between January 1997 and January 1999 with the UK grocery retailing group Tesco. The firm had established itself as the leading grocery retailer in the country with 24.2% of the grocery market share in 1998. During the same period it generated £16 billion of annual net sales from 1.56 million sq. metres (14.5 million sq. feet) of sales area at 568 outlets (Anon., 1999a). It was also the leading e-tailer and most profitable grocery retailing multiple in Europe, producing 3.25% return on sales (Kuipers, 1999).

Most of the fieldwork was conducted at the firm's commercial headquarters at Cheshunt. However, it also encompassed data from a regional distribution centre (RDC) and a retail store. Data were similarly collected from suppliers of product and packaging development services. These included firms that were third party contract manufacturers of products, as well as design studios, reprographic agencies and printing firms.

The Tesco commercial headquarters housed an administrative, office-based operation comprising a number of decentralised trading teams that were each responsible for one or more sub-category of products. Consequently, there were a number of trading teams operating in each product category, all of which reported to a category manager who was responsible for co-ordinating their activity. The commercial support functions, such as trading law, design and finance, were centralised and were not dedicated to serving any specific trading team.

Each trading team was composed of one or more buyer, food technologist and product development manager. However, the number of commercial staff allocated to the team in each of these roles depended upon the size of the product range (number of product lines) and level of product development activity within that team's remit. These staff were specifically responsible for sourcing new suppliers and products, negotiating commercial terms and conditions, developing new products and promoting them.

The trading teams used two generic processes to develop all of the firm's Value, Standard and Finest range private-label products. In 1998, these ranges represented 39.3% of the firm's total grocery sales. After Sainsbury (47.5% including Savacentre), this was the second highest private-label market share of any UK grocer (Anon., 1999b). The scope of these two generic processes encompassed the conception of the product through to its delivery at the retail store. One was used for the development of all the firm's food products, whilst the other was used for the development of all its non-food products. These same processes were also used for the introduction into the Tesco system of new branded products that had been developed by food manufacturing firms. Both of these generic processes were formally documented and used by the firm in a standard way. This case study discusses the food product development process.

Product development project types

Figure 10.1 summarises the different types of private-label product development projects that were implemented using the Tesco food product development process. These are presented in high to low innovation level sequence as recognised by the firm.

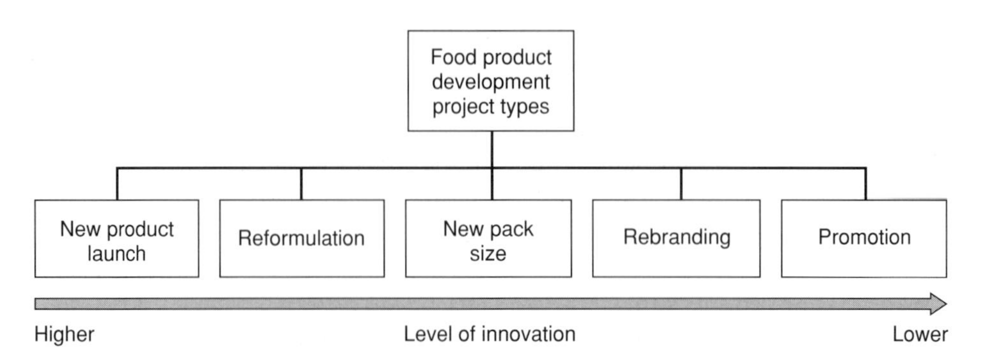

Figure 10.1 Tesco's food product development project types (adapted from company data).

The type of project with the highest innovation level was the new product launch. Unlike the other types which involved modifying an existing product, this involved developing a new product. Some of these were truly new (Ernst & Young/ACNielsen, 1999), meaning that they had not been seen before and brought new value or a new consumption experience to the consumer; others involved developing a me-too, which imitated the characteristics of an existing retailer private-label or manufacturer branded product.

The second type of development project was the reformulation, in which the recipe of an existing product was reformulated either to improve its quality or reduce its cost. Reformulation also involved changing its packaging text (words) to reflect the new recipe.

The third type of project was the new pack size, which included developments such as the creation of a family rather than individual sized portions of an existing private-label product. A new pack size also involved changing the product recipe and therefore again required a packaging change.

The penultimate type of project was rebranding, which entailed changing the packaging material without changing the product recipe itself. Typically, a buyer initiated this type of project to refresh the product's market appeal.

The final type was a promotion, which was a temporarily marketed product that typically involved the development of price flashed packaging. A price flash was a standard price incorporated as part of the packaging design and printed in a prominent position on the front of the packaging.

Process description

The standard Tesco food product development process had a cycle time (total elapsed time between the start and end stages of the process) of 30 weeks. Figure 10.2 is a schematic representation of this process which provides a vehicle for explaining the main activities that were conducted during each of its stages.

Planning

The planning phase comprised two product-independent stages. The first of these involved constructing the food trading plan. This document was an output of the retailer's annual strategic planning exercise and defined the performance objectives for all the firm's food product categories. In accord with the efficient consumer response (ECR) category management process (Andersen Consulting, 2000), the second planning stage entailed developing a separate category plan for each food product category. This plan defined the category performance objectives that were designed to contribute to the higher level performance objectives in the food trading plan. Category plan production required the trading teams to agree development priorities concerning which new product ranges to develop and which existing product ranges to extend. This exercise culminated in the production of the product development list giving each planned product development project in launch date sequence.

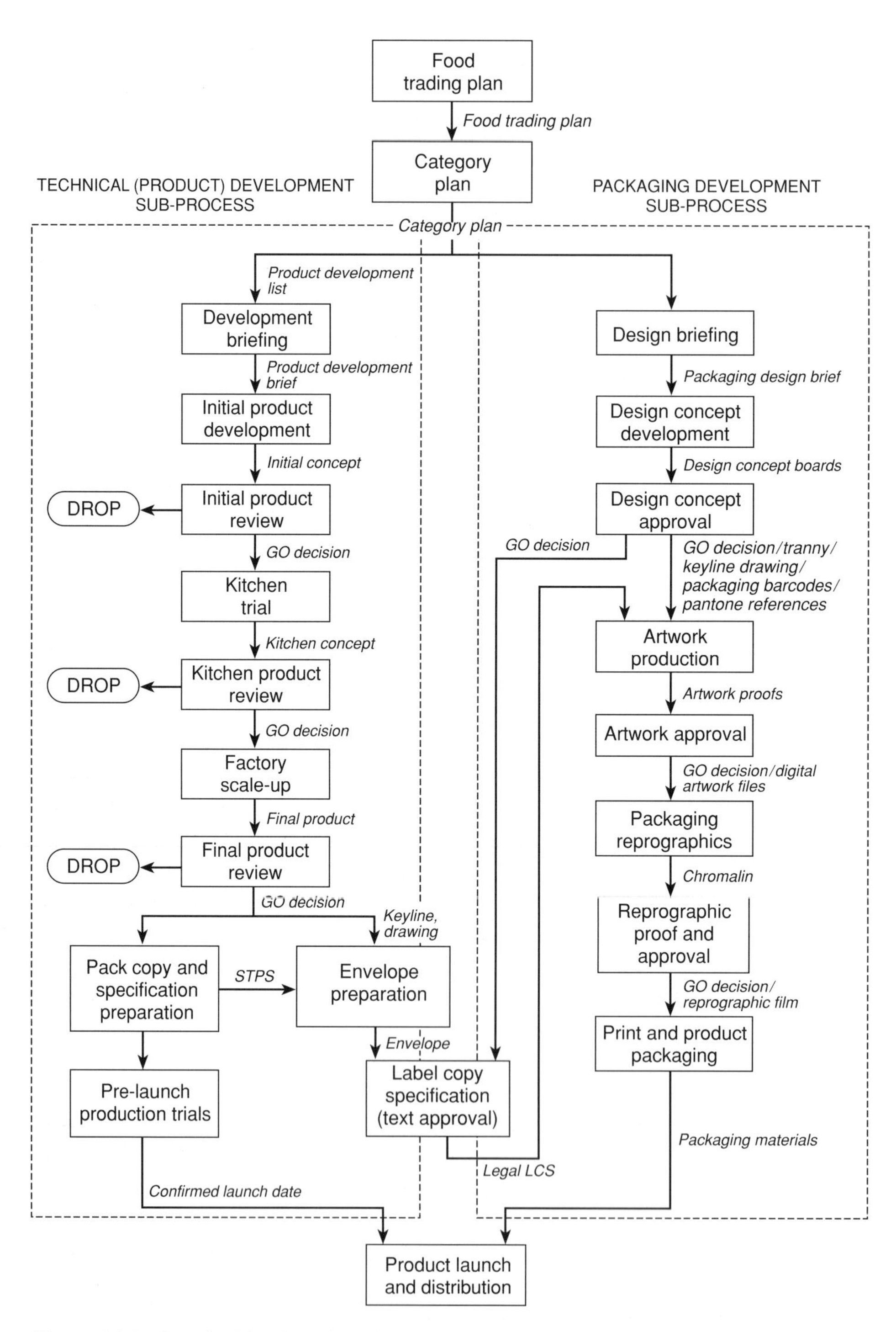

Figure 10.2 Standard food product development process (adapted from Tesco data) (*STPS* supplier technical product specification).

The technical (product) development sub-process

Technical (product) development encompassed all the stages required for developing the product recipe. Taking in sequence each of the products itemised in the product development list, the first stage was the development briefing, which required the trading team to produce a product development brief describing the product concept and including details such as the legal title, target price, weight, shelf life and launch date of the new product; it also included a technical description of its desired flavour, appearance, texture and aroma. The trading team then arranged a meeting with the selected contract manufacturer to discuss the concept and to agree a time-scale for the completion of each of the stages in this sub-process.

The contract manufacturer subsequently conducted any market research, comparative shopping (to assess competitive products), food shows and restaurant visits that were required to elaborate upon the concept about which they had been briefed, after which they developed an initial product concept based upon their findings. The product was reviewed by a Tesco food technologist, who made a decision to either drop (kill) the project or go (proceed) to the next stage in the process; such a review is referred to as a stage gate (Cooper, 1988).

If a decision was made to go, the initial product concept entered the kitchen trial stage during which the manufacturer's product development team worked on domestic-scale kitchen equipment to develop a kitchen concept that had acceptable eating and visual qualities.

If a decision was again made to proceed with the project, the process entered the factory scale-up stage, which was again conducted at the product manufacturer's factory. The purpose was to scale-up the kitchen concept onto the manufacturer's industrial-scale production equipment whilst matching the visual and taste standards established during the previous stage. Three tests could be conducted at this point: the first was a microbiological test to confirm the product's shelf life; the second was a home freezing test to establish whether it was suitable for freezing; the third was a travel test to ensure the product did not degrade microbiologically when transported between the factory and the consumer's home. The production costs were then finalised.

The product now entered the final product review stage. The trading team formed a taste panel and assessed the product's appearance, aroma, flavour and texture, again resulting in either a drop or go decision. If the decision was made to proceed, the process entered the pack copy and specification preparation stage. The manufacturer's product development team was required to use a computer document template supplied by Tesco entitled the supplier technical product specification (STPS) to detail the full technical specification of the new product, including the list of ingredients and nutritional content to be printed on the packaging material together with recipe, food additive and food intolerance information, the last being provided for the benefit of consumers with food allergies to indicate whether the production process and ingredients could be guaranteed free of traces of potentially harmful substances such as gluten and nuts. A hard (paper) copy of the STPS was printed off and posted to the trading team for review. If it did not contain any errors, it was placed in an envelope with the three other documents required by the Tesco Trading Law Department during the text copy and approval stage. The first of these detailed the marketing text copy that was to appear on the packaging and included the contact information, product title and claims such as healthy eating and suitable for vegetarians. The second document described the cooking instructions and serving suggestions that were to be included in a panel on the back of the packaging. The final document was a copy of the keyline drawing which provided technical information on the product manufacturer's production machinery. The studio commissioned to produce the packaging design used this information to establish the dimensions within which they could locate the artwork and text copy; the printer also used it. This was to ensure that the packaging material that was produced could be used on the manufacturer's packing equipment.

The receipt of the envelope by the Trading Law Department marked the start of the label copy specification (text) approval stage. First, the Trading Law Manager transcribed the information contained therein to the Tesco computer system. Some of this information was used to update the computerised databases maintained by the department including the food intolerance, additives and ingredients databases. The manager next assessed the legality of the labelling information derived from the STPS and amended this as necessary, finally producing a document entitled the legal label copy specification (LCS) detailing the legally approved packaging text.

The pre-launch trials were conducted in parallel with the development of the packaging. These were a rehearsal by the manufacturer for a full-scale production run of the developed product, which enabled the launch date to be finalised. Blank packaging was used because the final packaging material was still being developed at this point.

It is worth noting that the product development process did not involve any pre-launch consumer trial or market testing stage to assess the level of likely market acceptance of the prototype private-label product. Such a stage is common in a development process of a consumer durable product such as a car (Clark & Fujimoto, 1991), or for expensive industrial goods and products incorporating new technologies (Kotler, 1994). However, the vast majority of food products are me-too (derivative) in nature (Ernst & Young/ACNielsen, 1999). They are therefore relatively cheap to develop because the research and development and advertising costs have already been borne by the products upon which they are based. Taken in conjunction with the number of stage gate (Cooper, 1988) reviews by technical staff to which such new food products are subjected, they consequently carry a low risk of failing to recoup their development costs. Market testing for such a me-too private-label food product is not therefore perceived to be a cost-effective activity.

The packaging development sub-process

Packaging development encompassed all the stages required for developing the packaging into which the new product was to be inserted. The first stage was the design briefing which involved a meeting between the trading team and a design executive from the Tesco Design Department who was responsible for the packaging artwork for that product sub-category. The design executive was briefed about the nature of the new product and the product manufacturer involved, after which deadlines for the delivery of the packaging components of the new product were agreed.

Using the information obtained during the design briefing, the packaging design brief was created by the design executive. This formed the requirement specification for the packaging design and had two parts: the first was the marketing element and included items such as the character of the product and brand, marketing objectives and the target completion dates for the design stages; the second included items such as the key visual signals, photography and illustration requirements, and an indication of the structural components of the pack. The design executive then organised a meeting with the design studio and reprographic agency to brief them about Tesco's packaging requirements for the project. The design studio developed a number of design concept boards using this information. Each of these was a full-scale colour mock up of the final packaging design on an A3 board; they were intended to convey the 'mood' of the packaging design.

The sub-process now entered the design concept approval stage during which the design studio organised a short meeting with the Tesco design executive and buyer to present the design concept boards so that one of them might be selected for development. When a design concept board had been approved and the legal LCS was available, the artwork production stage started. A copy of the legal LCS was sent to the design studio and the product manufacturer simultaneously sent preproduction samples of the product to be photographed to ensure that an accurate illustration was reproduced on the front of the packaging. The design studio used a number of inputs to produce the packaging artwork, including the transparency positive of the photograph supplied by the photographer, the legal LCS, keyline drawing, packaging barcodes and colour reference documents supplied by Tesco. The resulting artwork proof was a two-dimensional representation of the product packaging; it included the exact location, size, orientation and colour of all the pictures, illustrations, graphics and text to appear on that packet, together with information on the position of the fold lines and certain printing machine calibration marks.

Copies of the artwork proof were sent to the Tesco Design Department, which on receipt started the Artwork Approval stage of the process, and to the Trading Law Department, where it was reviewed in conjunction with the earlier legal LCS document to ensure that the text and artwork for the project complied with labelling legislation when they were considered together. On approval, the design studio was contacted to authorise transmission of the digital artwork files to the reprographic agency, where its receipt started the packaging reprographics stage. This involved conversion of the digital artwork into a format suitable for physical printing. This process gave an exact, colour copy of the final packaging material laid out flat on an A3 board which was sent to the retailer to be checked for accuracy; the reprographic agency was informed when this was approved. This acted as a signal for them to produce a high quality reprographic film of the packaging design to be sent to the printer to start the print and produce packaging stage. At this point the printer contacted the product manufacturer to confirm the scheduled product launch date, and then produced an initial production batch of packaging material.

Product launch and distribution

Finally, during product launch and distribution, the batch of packaging material was delivered to the product manufacturer in advance of the scheduled launch date. It was then used to pack the newly manufactured product, which was delivered to the Tesco RDC network on the day of the launch.

Whilst the printer was planning the production of the packaging material, a member of the trading team updated the central Tesco stock control computer system with details of the packaging barcodes for the new product to enable it to be received at the firm's RDCs and retail stores. The computer update also acted as a signal to the trading team to inform store managers of the impending product launch by transmitting its details via an internal bulletin document.

Volume of product development activity

Changes in consumer behaviour within the UK food market over recent years can be explained by economic, demographic and consumer concern factors (Traill, 1998). These changes seem to have driven increased levels of product development activity. For example, the greater participation of women in the paid workforce, growth in total household expenditure on food, and increased utilisation of appliances for the storage and preparation of food have influenced increased consumption of convenience and value-added food products. Product development activity has also been stimulated by the increased demand for diet and health foods, and an increase in the variety and technology of packaging materials.

This trend was reflected in the volume of annual product development activity recorded at Tesco, which was characterised as an 'avalanche' by one member of staff. In 1995–1996, over 2500 new products of all the types illustrated in Figure 10.1 were developed, with 50% of these being new product launches. With no significant increase in development staff resource, this increased to over 6000 new products in 1997–1998, over 50% again being new product launch types.

In 1999–2000, the level of annual development activity increased further, to over 8500 new products. Much of this increase was attributable to three product labelling initiatives that were implemented by the firm during this period. These manifested themselves as rebranding projects because they involved changing the packaging of existing products but not their underlying recipe. The first of these was quantitative ingredient declaration (QUID), a legislative requirement due to come into force on 14/02/2000. QUID was designed to help customer comparison between products. The legislation demanded that volume percentages be either included in the ingredient list or alternatively reflected in the legal name of the food product on the front of the packaging.

The second initiative was a competitive response to the retailer Co-op called honesty in labelling which involved providing a complete breakdown on the product packaging of the components of all the ingredients comprising more than 2% of the product's volume. The initiative also called for additional information to be displayed on the label, including country of origin, the country where it was packed, how long it could be stored after being opened, its nutrition per serving (not per 100 g) and also its salt content per serving. The final initiative was called GMO-free and involved clearly labelling all products that contained ingredients with genetically modified organisms (GMOs).

Although this increase in product development activity had been enabled by a great improvement in the productivity (quantity of work produced in the time available) of the staff involved in the process, there were still issues with the development process itself, one of the main ones being the quality of supplier technical product specification (STPS) documents submitted by the product manufacturers. It was reported that all these documents needed amendment by the Trading Law Department when they were checked at the text copy and approval stage of the process; this usually required a trading law manager to contact the manufacturer's product development team on one occasion or more to query and resolve the problems. It was also estimated that 10% of all STPS documents were beyond such amendment and had to be reworked and resubmitted by the manufacturer's product development team. This rework was disruptive and very costly to the retailer and manufacturer. It also typically added a number of weeks to the product development cycle time, which incurred the further (opportunity) cost penalty of lost sales.

A number of improvement initiatives were subsequently implemented by Tesco to overcome this STPS quality problem and to help improve the general quality, cost and cycle time performance of the food product development process, most of which were methodological in nature, involving a change to the organisation of the process and the working practices associated with it. However, one of the most important was technological in nature and involved the development of the Tesco Information Exchange (TIE).

Tesco Information Exchange (TIE) – an innovative application of information technology

Tesco had been collaborating with the UK software arm of General Electric (GE) to develop a secure extranet system called Tesco Information Exchange (TIE); this system was implemented in 1999. TIE provided a shared on-line environment that was accessible by web browser and which enabled the retailer and its selected private-label and branded suppliers to collaborate on the workflow of key business processes. When it was originally conceived, TIE was designed to enable the product promotion process to be managed more effectively with these suppliers. However, product development became the system's second key business process.

Being web-based, TIE was much cheaper to implement and maintain than electronic data interchange (EDI), which involved the electronic transfer of structured data (trading documents) between two computer systems using an agreed message protocol such as TRADACOMS. EDI meant that the customer and supplier had to invest in compatible hardware and software, and lease a telephone line for the digital communication link. TIE was also more flexible than EDI because it allowed a high degree of parameter tailoring. This tailoring extended to the software applications that could be embedded in the system, which was a feature that enabled private applications to be distinguished from community applications. Community applications could be used by all the firms that were connected to the TIE system; conversely, private applications could only be seen by the single retailer or supplier concerned.

The TIE system had three main areas. The first of these was the general information area that contained information such as retail store and staff contact details. The second area contained supply chain data, such as daily sales, service levels and stock levels. The final area contained the workflow tools for the promotions management and new product development processes discussed above.

The TIE product development workflow tool

The workflow tools presented on-line versions of the forms that were used during the process. They also provided guidance on how each form was to be completed, and on-line checks to ensure that form field (box) entries were valid. These forms were then presented in the correct (workflow) sequence to the party responsible for completing them at that stage in the process.

When it was implemented, the product development workflow tool simplified the product development process explained in the previous section. When the process entered the pack copy and specification preparation stage, the manufacturer's product development team called up a TIE version of the STPS form onto their computer screen, typed in the new product's technical details and then pressed the enter key to signal its completion. This made the STPS details available to all parties in real-time and automatically sent a message to the Tesco trading team and Trading Law Department to signal its availability for review.

This approach had direct cycle-time and cost benefits because there was no longer a need for the manufacturer to print off a hard copy of the STPS document and post it to the retailer. The TIE system also meant that it was no longer necessary for the Trading Law Manager to re-type the 30 pages of STPS data into the Tesco computer system during the text copy and approval stage of the process, because the data had already been entered. This eliminated further effort and the risk of data transcription error, along with their associated cycle-time and cost penalties.

Even greater benefits were provided in the guise of the validation rules that could be built into the TIE version of the STPS form. These rules ensured that the field entries made by the product manufacturers were consistent. Form completion was therefore standardised. The validation rules also prevented many common errors such as non-valid ingredient breakdowns and ingredient combinations from being entered into the computer system. This greatly reduced the volume of STPS related queries, amendments and rework being generated when the document was subsequently reviewed during text copy and approval. Such a preventative approach is endorsed by contemporary quality gurus such as Fiegenbaum (1983) and Deming (1988) as the most cost effective approach to quality management.

The final benefits that were attributable to the TIE product development workflow tool were associated with the centralised control that it provided. Apart from ensuring that a current version of each form template was always available for immediate use, the system maintained only a single copy of each completed form. This provided document control because it was not possible to have uncontrolled copies of these forms in circulation which had sometimes been a problem in the earlier process, resulting in further wasted effort and time.

When taken in conjunction with its accompanying supplier development initiative, the implementation of the TIE product development workflow tool was credited with a 20% reduction in the cycle time of the text copy and approval stage of the process and an equivalent reduction in the STPS rejection rate. Data on the cost savings and additional revenues associated with this performance improvement were not available; however, these must have been very significant given the volume of annual product development activity undertaken by Tesco.

Conclusions

It is possible to derive three key conclusions from this case. The first is the need to manage processes holistically if optimum process quality, cost and cycle time performance are to be achieved. This corroborates Goldratt and Cox's (1993) statement that,

> '. . . a system of local optimums is not an optimum system . . .'.

The second conclusion is that quality theory is just as applicable to the process of product development as it is to the process of acquiring and fulfilling customer orders, which is the main subject of supply chain management. It is therefore quality capable development processes that produce quality new products. Such quality capable processes can also be expected to consistently produce these new products at less cost and with a shorter cycle time because of their emphasis on the prevention of development errors rather than their 'cure' in the form of rework.

Finally, the case emphasises the need to reorganise and simplify the underlying process before applying technology to achieve further performance improvement. The supplier development initiative that was implemented in conjunction with the TIE product development workflow tool also highlights the benefit of complementing such technological implementations with supporting initiatives that are designed to maximise their utility. The case for putting 'methodology before technology' (Naim, 1996) is therefore as strong for product development as it is for the order acquisition and fulfilment processes.

Study questions

1. Consider three new products that you encountered when you last visited your local grocery store. How might these be classified according to the food product development project types illustrated in Figure 10.1?
2. Figure 10.2 illustrates the development process used by Tesco. Select a product from your refrigerator and consider how this product and its label information might have been developed using the described process. How long do you think it took to complete each stage of this process?
3. Again referring to a selection of products selected from your refrigerator, examine in detail the information contained on the product labels. What different claims are made on these? What do you think could be the consequences to the consumer and retailer if this information was inaccurate?
4. Referring to the literature on product development, what 'best practices' exist? How might these be applied to this case?
5. This case study explained the application of the TIE system within the food product development process. How might the utility of the TIE system be enhanced within this process? What other applications of technology might enhance the performance of this process?

References

Andersen Consulting (2000) *The Essential Guide to Day-to-Day Category Management*. ECR Europe, London.

Anon. (1999a) *Supermarkets and Superstores*. 16th edn. Market Report. Key Note Publications, London.

Anon. (1999b) *Own Brands*. Market Report. Key Note Publications, London.

Clark, K.B. & Fujimoto, T. (1991) *Product Development Performance: Strategy, Organization and Management in the World Auto Industry*. Harvard Business School Press, Boston.

Cooper, R.G. (1988) *Winning at New Products*. Kogan Page, London.

Deming, W.E. (1988) *Out of Crisis*. Cambridge University Press, Cambridge.

Ernst & Young/ACNielsen (1999) *Efficient Product Introductions: The Development of Value Creating Relationships*. ECR Europe, London.

Feigenbaum, A.V. (1983) *Total Quality Control*. McGraw-Hill, New York.

Goldratt, E.M. & Cox, J. (1993) *The Goal*. 2nd edn. p. 208. Gower, Aldershot.

Kotler, P. (1994) *Marketing Management: Analysis, Planning, Implementation and Control*, 8th edn. Prentice-Hall, New Jersey.

Kuipers, P. (1999) Top retailers: Walmart versus the rest, *Food International* **4** (October), 15–23.

Naim, M.M. (1996) Methodology before technology, *Manufacturing Engineer* **75**(3) 122–125.

Traill, W.B. (1998) Structural changes in the European food industry: consequences for competitiveness. In *Competitiveness in the Food Industry* (Traill, W.B. & Pitts, E., eds) pp. 35–57. Blackie Academic & Professional, London.

Chapter 11
Third Party Logistics in the Food Supply Chain

A.C. McKinnon

Objectives

(1) To examine the different roles that logistics service providers can play in the food supply chain.
(2) To review the recent growth of logistics outsourcing in the food sector.
(3) To assess the benefits that food producers and distributors derive from contracting out logistical activities.
(4) To explore the different outsourcing strategies that these companies can adopt.

Introduction

Much of the movement, storage and handling of food products is undertaken by logistics service providers (LSPs) on a third-party basis for food producers and distributors. Third party logistics has been defined as 'the use of external companies to perform logistics functions that have traditionally been performed within an organisation' (Lieb *et al.*, 1993). Over the past quarter of a century, an increasing proportion of logistics expenditure has been outsourced to these 'external' LSPs, making them major players in the food supply chain. This process has not been confined to the food industry. It is part of a worldwide, cross-sectoral trend towards greater concentration on core business strengths and the externalisation of ancillary activities that are not considered to be a source of competitive advantage. As many companies regard logistics as 'ancillary' rather than 'core' it has been an obvious candidate for outsourcing. The nature and extent of this process has, however, varied among sectors. In this chapter we shall focus on the outsourcing of logistics in the food industry.

The role of third party logistics service providers

LSPs can perform a range of logistical services at different levels in the food supply chain. Their varying roles can be classified in several ways.

Table 11.1 Major UK logistics service providers distributing food products (Distribution Business, 2002).

Company	Turnover (£m)	Total staff	Warehouse space ('000 sq. metres)	Vehicles (in UK)
Christian Salvesen	830	16 000	1 396	2 130
Exel	4 500	60 000	—	—
Frigoscandia	—	550	1.19 million m^3	350
Gist	231	4 500	—	532
Hays	1 000	14 000	1 035	1 200
NFT Distribution	100	1 520	50	500
TDG	530	8 500	1 000	2 150
Tibbett & Britten	1 970	36 200	1 457	1 448
Wincanton	770	16 000	826	3 900

Function

Logistics comprises several inter-related activities, mainly freight transport, warehousing, inventory management, materials handling and related information processing. Companies can outsource these activities individually or in various combinations. For example, many food manufacturers contract out the trunk movement of finished product from factories to retail distribution centres, essentially buying a basic transport service. Supermarket chains, on the other hand, typically employ contractors to run their distribution centres, manage the stock replenishment process and provide store delivery. Functional variability exists on the supply side as well as the demand side. The logistics market is dominated by small enterprises which provide a transport or storage service in isolation. At the other end of the spectrum are large logistics contractors capable of providing a full range of distribution services across a wide geographical area. Table 11.1 provides summary statistics on major UK LSPs that are heavily involved in the distribution of food products.

LSPs have steadily expanded the range of activities they perform and placed greater emphasis on the value that they can add to a client's products and services. Many now include tracking and tracing, the handling of waste and various forms of product customisation in their service portfolio. Recent studies have forecast that this functional diversification will continue for the foreseeable future (McKinnon & Forster, 2000). LSPs are also assuming a more prominent role in the design of logistical systems, often undertaking logistics planning as part of the tendering process.

Over the past five years, a new type of logistics service has evolved, providing higher level strategic support for companies wishing to integrate their supply chains. The consultancy company Accenture has applied the term Fourth Party Logistics™ (4PL) provider to a 'supply chain integrator that assembles and manages the resources, capabilities and technology of its own organisation and those of complementary service providers to deliver a comprehensive supply chain solution' (Bauknight & Bade, 1998). This generally involves the formation of an alliance of

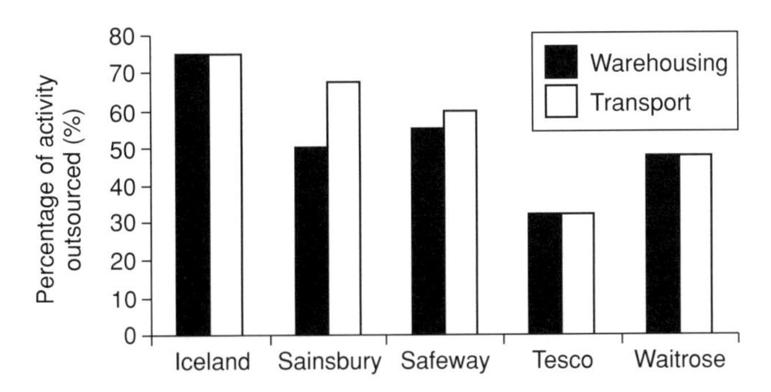

Figure 11.1 Variations in the level of outsourcing by major UK supermarket chains (Source: Marchant *et al.*, 2000).

consultancy, IT, financial and logistics businesses. The co-ordinating agency (or 4PL) 'need not have any specific resources, capabilities or technology itself, so long as it has the ability to analyse the client's requirements and take contractual responsibility for delivering to the client an optimised supply chain solution built from the integrated capabilities and resources of its partners' (Bedeman, 2001). Very few working examples of 4PL have emerged so far and, to the author's knowledge, there are none as yet in the UK food sector.

Degree of outsourcing

Some companies are prepared to outsource their entire logistics operation, while others prefer to retain a significant in-house (or 'own account') operation. Figure 11.1 shows variations in the proportion of transport and warehousing expenditure outsourced by major supermarket chains in the UK. Partially outsourcing the logistics function offers a number of benefits: it ensures that the company maintains management expertise in the field, spreads the risk of service disruption (possibly as a result of labour disputes) and permits benchmarking of efficiency and service standards. This casts the LSP in a supportive role, sometimes in a geographical sense extending the reach of a company's distribution system into more remote and less populated areas. Relative dependence on LSPs can also vary through time, where companies experiencing wide seasonal fluctuations in supply and/or demand must employ contractors to provide additional transport and storage capacity at peak times. This particularly applies to the food industry where there are wide seasonal variations in the input of agricultural produce and in consumer demand for many foodstuffs.

Level in the supply chain

As shown in Figure 11.2, the grocery supply chain can be divided into three sections. Primary distribution is the bulk movement of processed food from factories to regional distribution centres (RDCs). Secondary distribution is the delivery of

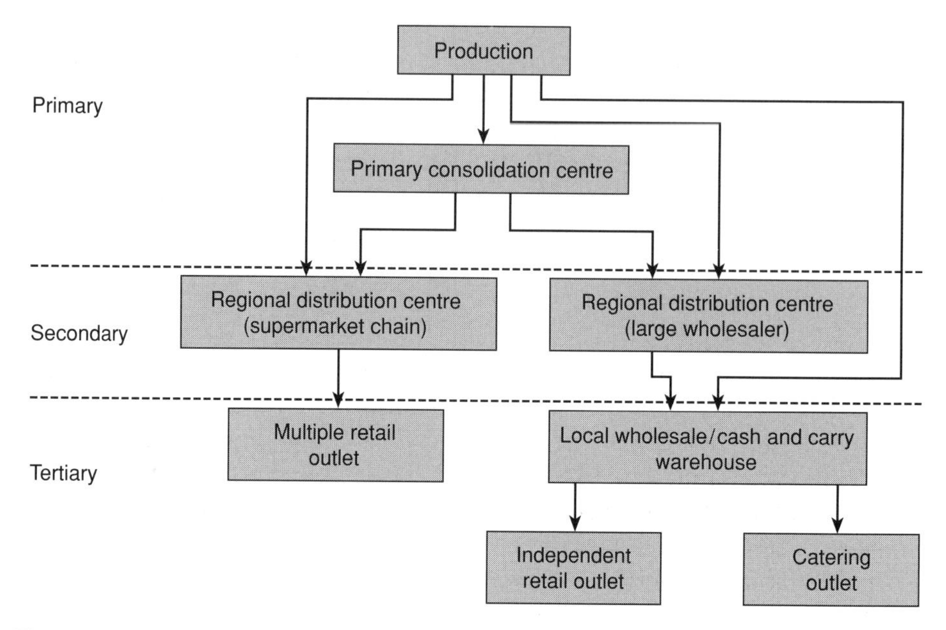

Figure 11.2 Structure of physical distribution channels in the food sector.

mixed loads from multiple retailers' RDCs to their shops. The term tertiary is applied to local deliveries from wholesale warehouse to independent retailers and catering outlets. In the UK, third-party penetration is greatest at the primary and secondary levels.

Primary level

At the primary level, LSPs can fulfil several roles:

(1) *Trunk haulage*: This is bulk movement of supplies, generally in full truckloads, from the producer's factory or warehouse to the RDCs of retail or wholesale customers.
(2) *Primary consolidation*: Many retail and wholesale orders, even those destined for RDCs, are of less-than-truckload quantities. Delivering these orders directly to RDCs is inefficient and creates 'backdoor congestion' problems for the customers. By channelling these orders through primary consolidation centres (PCCs) it is possible to cut transport costs and ease the pressure on reception facilities at RDCs. Most of these consolidation centres are operated by third-party logistics providers on a shared-user basis. In some cases, the manufacturer arranges delivery to the PCC; in others the logistics provider collects the supplies from the manufacturers' premises. Retailers often nominate the carrier to be used for primary consolidation, though the related costs are generally incurred by the manufacturer. The demand for primary consolidation services has increased sharply over the past 10–15 years as quick response

replenishment has become the norm across the grocery supply chain (Fernie, 1994). In an effort to cut inventory levels, retailers have increased order frequency and reduced average order size. This was illustrated by a survey of the distribution operations of 42 frozen food manufacturers undertaken in the UK (McKinnon & Campbell, 1998). It is in the frozen food sector, where transport and storage costs are relatively high, that third-party primary consolidation has become most firmly established (Lindfield, 1998). Increasing amounts of chilled and ambient product are also being channelled through third-party consolidation networks.

(3) *Pallet-load services*: The volume of food distributed through pallet-load networks has also been expanding. These networks generally have a hub-satellite structure, similar to those of parcel carriers. They are designed to provide economical, overnight distribution of individual pallet-loads at a national level. Unlike the primary consolidation services which specialise in the distribution of food, the pallet-load networks handle a diverse range of products, mixing consignments of food with those of other products to achieve economies of scale in both transport and handling operations.

Secondary level

The involvement of LSPs at the secondary distribution level can also take different forms:

(1) *Integrated contract distribution*: This has been the most common form of outsourcing by large food retailers. Contractors are engaged to operate RDCs and provide store delivery within the surrounding area. Just under half of the RDCs used by major UK supermarket chains in 2000 were operated on a third party basis, the vast majority by the five major LSPs (Christian Salvesen, Exel, Hays, Tibbet & Britten and Wincanton) (Marchant *et al.*, 2000). Browne & Allen (1998) observe that 'the storage facilities owned and operated in-house by retailers tend to be the older, smaller facilities in their portfolio'.

(2) *Separate transport and warehousing services*: Some grocery retailers employ different LSPs to run their warehouses and provide store delivery. In some European countries this has been quite a common strategy. In the UK, however, it is a more recent development, prompted by a desire to achieve higher levels of vehicle utilisation. Traditionally, vehicles were dedicated to a particular RDC and only used to deliver to shops within its hinterland. Over the past decade, supermarket chains have been integrating their primary and secondary distribution operations, using delivery vehicles to collect inbound supplies and cross-haul inventory between RDCs (Marchant *et al.*, 2000). Trucks can now migrate across the distribution network, picking up more backloads and being more intensively used over the 24-hour cycle. This can lead to the adoption of a 'one fleet' approach to transport where the management of vehicle assets is centralised rather than devolved to individual RDCs. This

severs the link between warehousing and transport at the local level and makes it more appropriate to outsource the two functions separately. This is a strategy that the supermarket chains such as Sainsbury and Budgens have adopted. The extension of retailers' control over the inbound flow of supplies into RDCs has also enabled LSPs that were originally contracted to provide store delivery to assume a greater role in the primary movement of food products.

(3) *Reverse logistics*: Tightening packaging waste regulations has forced grocery retailers to develop systems for the return of waste material for recycling and handling units for re-use. Some supermarket chains have integrated these reverse logistics operations into their existing systems, establishing 'recovery units' at existing RDCs. Others have employed LSPs to develop separate waste collection and recovery networks.

Tertiary level

Only a small proportion of food deliveries at the tertiary level is outsourced. The companies providing these deliveries to independent retailers and catering outlets, mainly wholesalers and food service companies, generally regard local transport as a core activity involving close interaction with customers and heavy emphasis on service quality.

Degree of service exclusivity

Much of the outsourcing of food distribution in the 1970s and 1980s, particularly by retailers, was on a dedicated basis. All the logistical assets were assigned for the specific use of individual clients. This type of outsourcing might be considered 'quasi-own account' as it closely resembled in-house operations and denied LSPs the opportunity to improve asset utilisation by combining different companies' logistical demands. Dedication brought higher levels of service quality and control, but at premium rates. In recent years, many companies have been prepared to relax their earlier insistence on dedicated services in an effort to cut costs. Contractors can now carry traffic for other companies, even on client-liveried vehicles, generating additional revenue against which some of the costs of the contract distribution operation can be defrayed. At the other end of the spectrum, there has also been a healthy growth in demand for shared user (or 'groupage') services, such as those described above under the primary distribution heading.

Type of product

Within the food sector there is a diverse range of products with differing storage and handling requirements. Some LSPs can accommodate most, if not all, of these requirements, while others specialise in the distribution of particular types of food. For logistical purposes, food is usually classified in two ways:

(1) *Degree of temperature control*: Most food products are stored and transported at one of four temperature levels:
- –25°C for frozen products;
- +2°C for chilled products, such as dairy products and meat;
- +5°C for produce, such as fruit and vegetables;
- ambient temperature for products not requiring temperature control.

All the major British LSPs listed in Table 11.1 can provide multi-temperature distribution.

(2) *Handling characteristics*: The main distinction is between bulk products which are typically moved in tankers and hopper vehicles, and packaged goods which are generally transported in box vans or curtain-siders. Most packaged food products are handled in unitised loads such as wooden pallets and roll cages.

Division of asset ownership

A fundamental distinction exists between asset-based and management-only logistics contracts. The former require the LSP to invest in logistical assets. This investment is often confined to vehicles, which can be depreciated over a 4–5 year period. Where it also entails investment in property, such as a distribution centre, the client is usually bound by a much longer-term contract. Management-only contracts, on the other hand, leave the ownership of all the logistics assets in the hands of the client. The LSP assumes responsibility for recruiting and managing the staff and operating the assets. This type of contract is now very common in the grocery supply chain, particularly among the large supermarket chains. It gives clients greater flexibility and allows them to terminate contracts at relatively short notice.

Growth of logistics outsourcing in the food sector

In the UK, it is possible to chart the outsourcing of the transport function using data from the government's continuing survey of road goods transport (Department for Transport, 2002). This distinguishes 'mainly own account' transport from 'mainly public haulage' and disaggregates the freight statistics by commodity type. The proportion of tonne-km carried by public hauliers provides a good indication of the level of transport outsourcing. For food products as a whole, this proportion rose from 48% in 1991 to 61% in 2001 (Figure 11.3). The level of outsourcing is significantly higher for agricultural produce than for the 'other foods' category, which comprises mainly processed foods, though the latter has exhibited a faster growth of outsourcing over the past decade. Although the food sector overall has shown a slightly lower propensity to externalise the transport function than the average for other sectors, this differential has been narrowing in recent years.

No comparable statistics are available on the outsourcing of other logistical activities for the food sector as a whole. The outsourcing of transport and warehousing by supermarket chains has, however, been monitored by the Institute of Grocery Distribution (IGD) (Marchant *et al.*, 2000) by measuring the level of

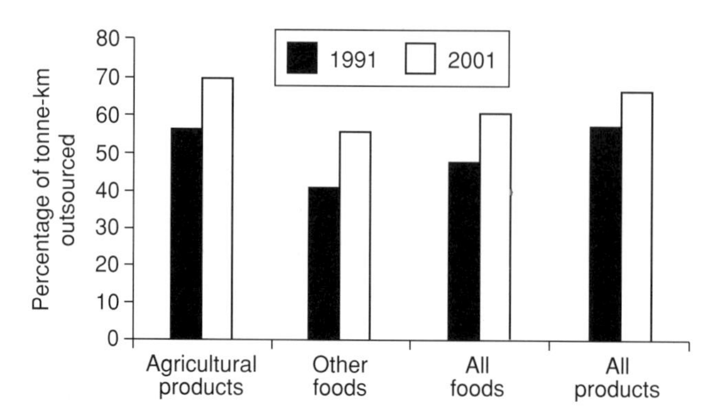

Figure 11.3 Proportion of freight movement outsourced in the UK in 1991 and 2001 (Source: Department for Transport, 2002).

outsourcing in terms of expenditure rather than traffic levels. The IGD statistics indicate that, following a growth of outsourcing in the late 1970s and 1980s, the level has fluctuated around a fairly stable mean during most of the 1990s. It was acknowledged in the early 1990s that contract distribution in the grocery sector (Cooper & Johnstone, 1990), particularly by supermarket retailers, had reached 'saturation level'; the IGD surveys conducted over the past decade broadly support this view.

The third party logistics market has, nevertheless, continued to grow in the retail grocery sector, as a result of three trends:

(1) Expansion of the total grocery market.
(2) Increase in multiple retailers' share of this market.
(3) Diversification of the range of services offered by LSPs.

The demand for third party logistics services from food manufacturers has also continued to grow. This has been supplemented by the emergence of new market opportunities for LSPs in, for example, the primary distribution of wholesale supplies to cash and carry depots, grocery deliveries to petrol forecourt outlets and the delivery of catering supplies to hospitals.

The trend to outsource the physical distribution of food products would not have been so strong and enduring had it not yielded major benefits. In the next section we examine these benefits and weigh them against some of the potential disadvantages.

Benefits of outsourcing logistical activities

Several studies have explored companies' motives for outsourcing logistics (e.g. Fernie, 1990; Peters *et al.*, 1998; Razzaque & Sheng, 1998; Browne & Allen, 2001). Many companies outsource logistical activities mainly to cut cost. McIvor (2000) argues that, 'Outsourcing decisions are rarely taken within a thoroughly strategic perspective with many firms adopting a short term perspective and being motivated primarily by the search for short term cost reductions'. Third party operators can offer these cost reductions for several reasons. Where they are able to consolidate several clients' traffic in groupage or network services, they can achieve higher levels of asset utilisation. They can balance out the fluctuations in traffic levels that individual clients experience, maintaining higher load factors in vehicles and warehouses than the typical own-account operator can achieve. Even where the third party service is provided on a dedicated basis, however, the contractor can still offer a cost advantage as a result of the bulk buying of equipment, fuel and materials, more effective management of the distribution operation and, often, lower labour costs. It is common for wage rates in in-house logistics operations to be significantly higher than those paid by outside contractors. Dedicated distribution operations are often contracted out on an 'open book', giving the clients a detailed break-down of the actual costs incurred and permitting comparison with own-account costs.

Outsourcing can also change the way in which logistical activities are financed. Where a contract is asset-based, the client can avoid investing capital in depots, vehicles and equipment, and treat logistics purely as a current cost element in the balance sheet. It can then concentrate investment in core activities, which generally earn higher rates of return, and gain the flexibility to 'contract for the necessary level of service to meet current demand' (Razzaque & Sheng, 1998). Outsourcing can also release capital already invested in an in-house logistics system. To facilitate the transition to a third party operation some contractors are prepared to acquire in-house logistics assets and re-employ some of the staff.

Outsourcing allows companies to minimise the risk associated with the ownership of logistics assets. Rapid changes in technology, regulations and market conditions introduce a great deal of uncertainty into the management of logistics. As functional specialists, LSPs should be better able to deal with this uncertainty than in-house logistics management.

Much of the literature emphasises the importance of asset provision in the decision to externalise or internalise the logistics function (e.g. Aertsen, 1993; Bourlakis, 1998). This reflects the heavy reliance on Williamson's transaction cost analysis (Williamson, 1979) as the dominant paradigm for research on logistics outsourcing, according to which the degree of asset specificity is a key determinant of the level of outsourcing. If assets have to be closely tailored to the needs of a particular company, it is more likely that the function will be internalised. Low asset-specificity, on the other hand, would be associated with a stronger tendency to outsource. As explained, however, much of the contract distribution of food products is not asset-based. Food manufacturers and retailers can invest in customised logistical systems and merely employ outside agencies to operate them. The traditional argument that companies outsource to avoid committing capital to logistics is much less relevant today in large sections of the food industry. As new multi-temperature composite distribution centres in the grocery sector cost around £30–40 million to build (Stephens & Wright, 2002), few LSPs would have the financial resources to provide dedicated logistics services on an asset-ownership basis.

Many companies also hope that by outsourcing they can upgrade the standard of logistical service. Few empirical data are available to assess the extent to which this hope is actually realised. Any service improvement will be relative to the standard previously achieved by the in-house operation and this will vary from company to company. There are circumstances in which the use of third party logistics will offer clear service advantages. By consolidating several companies' traffic, for instance, an LSP can justify making more frequent deliveries to customers than individual own-account operators, particularly in more remote areas. Logistics providers can also achieve higher service standards through the use of more sophisticated IT systems and more advanced equipment.

Ironically, concern for service quality discourages some companies from outsourcing. A survey by Croucher (1999) found that the two most important reasons cited by 'insourcing' companies for not contracting out were that, 'respondents were not convinced that the contractors could provide a better service than the company itself' and that they 'saw their drivers as important ambassadors of their business, critical to high standards of customer service'.

For these reasons many companies have been reluctant to relinquish direct control of their logistics operations. Advances in IT, however, have made it possible to monitor and control third party operations as tightly as in-house systems. This addresses another issue that Williamson (1979) identified as strongly influencing the outsourcing decision, namely the ease of measuring the performance of the contract service. Most large distribution contracts are now rigorously appraised to ensure that service levels meet or exceed the agreed standard. Indeed, some companies have found that the 'transparency' of their logistics performance has improved following outsourcing to an LSP.

Several surveys have, nevertheless, detected a significant level of dissatisfaction with the performance of LSPs, particularly in the later stages of a contract period. In the early stages, clients usually enjoy an initial stream of cost savings, sometimes supplemented by various service improvements. Once these are exhausted, enthusiasm for the 'third party solution' sometimes wanes. A common complaint is that LSPs do not do enough to secure a continuing flow of efficiency gains and service improvements. As PE Consulting (1996) observed, 'almost two in three providers are believed to be essentially reactive in their approach. Customers are highly critical of the passivity of their providers. Customers are looking to providers to challenge them to introduce best practice and to find new ways to add value'.

In some cases, however, the contract is so tightly specified that the third party operator has limited scope to innovate. Clients must shoulder some of the blame for not establishing more open relationships with their LSPs and involving them more fully in operational and strategic reviews.

A good example of beneficial collaboration between a food company and LSP is provided by British Sugar and Tankfreight. These companies jointly designed a vehicle capable of transporting bulk sugar in one direction and packaged sugar in the other, greatly increasing opportunities for backloading. This would not have been possible without 'a close working relationship and trust between these organisations' (Browne & Allen, 1998).

In recent years, the main focus of research in this field has shifted from the initial outsourcing decision to the management of the client–LSP relationship (McKinnon, 1999). This is influenced by a range of factors, including the contract specification, the nature of client–contractor communications, and the complexity of the logistics task. The basis of the relationship is also determined at the outset by the nature of the outsourcing strategy that the company adopts.

Outsourcing strategies in the food sector

There are examples of firms in the food sector outsourcing logistical activities in many different ways. These various outsourcing strategies have been grouped into five general categories and ranked in terms of the strength of the initial association between client and LSP.

(1) *Spot hire of transport services*: Much of the trunk haulage of food products is purchased on a 'transactional' basis from carriers. The transaction relates to the movement of a single consignment and is not governed by a contract as such. This is usually considered a standard 'commodity' service to be purchased at minimum cost in the general haulage market. As this is a highly fragmented, intensely competitive market, companies can 'shop around' for the best rates on particular routes. Large food manufacturers with extensive distribution operations might regularly use hundreds of hauliers to handle primary movements between factories and distribution centres. Buying haulage services in this way minimises transport costs, but burdens companies with high transaction costs. Over the past decade, most companies have been reducing these transaction costs in several ways: some have simply reduced the pool of approved hauliers from which they regularly buy transport; others have employed an LSP to purchase haulage services on their behalf. This so-called 'freight management' option involves the LSP acting as an intermediary and sub-contracting transport operations to a pre-selected group of hauliers. The recent development of online freight exchanges is offering companies a cheaper and more flexible means of exploiting the highly competitive conditions which prevail in the haulage market. This form of business to business (B2B) e-commerce creates electronic auctions for haulage services, allowing shippers to conduct a wide search for suitable carriers at minimal cost (Rowlands, 2000; Mansell, 2001). One of the main European freight exchanges, Freight-Traders, was set up by the Mars group in 1999 and handled 140 million euro of haulage business in its first year, much of it for food manufacturers.

(2) *Contracting out the distribution operation*: The outsourcing of an integrated distribution operation is invariably done on a contractual basis. As noted earlier, the length of the contract is generally a function of the level of contractor investment in client-specific assets. It has been common practice in the food sector for companies to close down their existing in-house operation and adopt a 'clean slate' approach to outsourcing. The decision to outsource is often made as part of a fundamental restructuring of a company's logistical system.

(3) *System buyout*: As discussed earlier, logistics outsourcing sometimes involves the transfer of in-house assets and employees to the LSP. Only the larger LSPs have the capital resources to fund such a buyout and this usually requires a medium to long term commitment from the client.
(4) *Joint venture*: This allows a company to retain direct involvement in logistics and commercialise specialist expertise acquired over many years by in-house managers. In partnership with an LSP, the company creates a new distribution service which not only handles its own traffic but provides logistical support to other, sometimes competing, companies. The best example of such a joint venture in the food and drink sector is that established by Exel, the UK's largest LSP, and the brewing firm Bass.
(5) *System spin-off*: Some food manufacturers have effectively converted their in-house distribution system into a third-party network, initially by giving it the status of a separate profit centre and then granting it the freedom to carry traffic for others. Following the deregulation of the UK haulage industry in the 1970s, it was relatively easy for British firms to 'spin-off' their logistics function in this way. One of the first food companies to do so was United Biscuits, whose distribution division began trading as UB Distribution Services in competition with other independent LSPs. This represents the culmination of a process which begins with an own account operator absorbing some third party traffic to improve asset utilisation and ends with them creating a separate logistics business. The process is more one of diversification than outsourcing and requires the company to have core skills in logistics management.

Conclusion

A large and increasing proportion of logistical activity in the food sector is being outsourced to LSPs. These companies have a strong presence at most levels of the food supply chain and can provide a flexible mix of services on either a dedicated or shared-user basis. By contracting out their logistics, food producers and distributors can both cut costs and improve service quality. The food sector, nevertheless, presents particular challenges for LSPs because of the perishability and fragility of many of the products, the time-sensitivity of deliveries, and the high variability in throughput levels resulting from seasonality and promotional activity. As there are few examples of logistical activities being brought back in-house, one can conclude that LSPs are generally managing to meet the exacting requirements of companies in this sector.

As the level of outsourcing in some parts of the food industry is nearing saturation level, LSPs are trying to expand by diversifying the range of services they provide and extending their operations geographically. There has also been much discussion in recent years about the possibility of a higher tier of 'asset-less' fourth party logistics providers gaining control of the design and management of supply chains. To date, however, there is little evidence of fourth party logistics operations affecting the distribution of food products. The nature of the relationship between

LSPs and their clients in the food sector is, nevertheless, evolving, with greater emphasis being placed on partnership, trust and transparency.

Study questions

1. Why has so much of the distribution of food products been entrusted to logistics service providers?
2. How does the nature of the logistical activities outsourced vary as one moves down the food supply chain from farm to factory to shop?
3. Imagine that you are the chief executive of a biscuit manufacturing company which wishes to reduce its dependence on in-house distribution facilities. What outsourcing options should you be considering?
4. Is the development of fourth party logistics likely to have a major impact on the food supply chain over the next decade?

References

Aertsen, F. (1993) Contracting out the physical distribution function: a trade-off between asset specificity and performance measurement. *International Journal of Physical Distribution and Logistics Management* **23**(1), 23–29.

Bauknight, D. & Bade, D. (1998) Fourth party logistics – breakthrough performance in supply chain outsourcing. *Supply Chain Management Review, Global Supplement*, Winter 1998. Reed Elsevier, New York.

Bedeman, M. (2001) Is 4 more than 3+1. *Logistics Europe* **9**(1).

Bourlakis, M. (1998) Transaction costs, internalisation and logistics: the case of European food retailing. *International Journal of Logistics: Research and Applications* **1**(3), 251–264.

Browne, M. & Allen, J. (1998) Logistics of food transport. In *Food Transportation* (Heap, R., Kierstan, M. & Ford, G., eds). Blackie, London.

Browne, M. & Allen, J. (2001) Logistics out-sourcing. In *Handbook of Logistics and Supply Chain Management* (Brewer, A.M., Button, K.J. & Hensher, D.A., eds), pp. 253–268. Pergamon, Oxford.

Cooper, J.C. & Johnstone, M. (1990) Dedicated contract distribution: an assessment of the UK market place. *International Journal of Physical Distribution and Logistics Management* **20**(1), 25–31.

Croucher, P. (1999) Insourcing. *Logistics Focus* **6**(2).

Department for Transport (2002) *Transport of Goods by Road in Great Britain*, Transport Statistics, London.

Fernie, J. (1990) Contract distribution in multiple retailing. *International Journal of Physical Distribution and Materials Management* **19**(7), 1–35.

Fernie, J. (1994) Quick response: an international perspective. *International Journal of Physical Distribution and Logistics Management* **24**(6), 38–46.

Lieb, R.C., Millen, R.A. & Wassenhove, L.V. (1993) Third party logistics services: a comparison of experienced American and European manufacturers. *International Journal of Physical Distribution and Logistics Management* **23**(6), 35–44.

Lindfield, G. (1998) Retail consolidation in frozen food. *Logistics Focus* **6**(1).

McIvor, R. (2000) A practical framework for understanding the outsourcing process. *Supply Chain Management* **5**(1), 22–36.

McKinnon, A.C. (1999) The outsourcing of logistical activities. In: *Global Logistics and Distribution Planning* (Water, D., ed.), pp. 215–234. Kogan Page, London.

McKinnon, A.C. & Campbell, J. (1998) *Quick response in the frozen food supply chain: the manufacturers' perspective*. Christian Salvesen Logistics research paper no. 2, Heriot-Watt University, Edinburgh.

McKinnon, A.C. & Forster, M. (2000) *European Logistical and Supply Chain Trends 1999–2005: Full Report of the Delphi 2005 Survey*. Logistics Research Centre, Heriot-Watt University, Edinburgh.

Mansell, G. (2001) The development of online freight markets. *Logistics and Transport Focus* **3**, 7.

Marchant, C., McKinnon, A.C. & Patel, T. (2000) *Retail Logistics 2000*. Institute of Grocery Distribution, Watford.

PE Consulting (1996) *The Changing Role of Third Party Logistics – Can the Customer Ever be Satisfied?* Institute of Logistics, Corby.

Peters, M., Lieb, R. & Randall, H.L. (1998) The use of third party logistics services by European industry. *Transport Logistics* **1**(3), 167–179.

Razzaque, M.A. & Sheng, C.C. (1998) Outsourcing of logistics functions: a literature survey. *International Journal of Physical Distribution and Logistics Management* **28**(2), 89–107.

Rowlands, P. (2000) Online freight exchanges. *e.logistics magazine*, May.

Skjott-Larsen, T. (2000) Third party logistics – from an inter-organisational point of view. *International Journal of Physical Distribution and Logistics Management* **30**(2), 112–127.

Stephens, C. & Wright, D. (2002) The contribution of physical distribution management to the competitive supply chain strategies of major UK food retailers. *International Journal of Logistics: Research and Application* **5**(1), 91–108.

Williamson, O.E. (1979) Transaction cost economics: the governance of contractual relations. *Journal of Law and Economics* **22**, 233–261.

Chapter 12

Temperature Controlled Supply Chains

D. Smith & L. Sparks

Objectives

(1) To define temperature controlled supply chains.
(2) To understand the importance of temperature controlled supply chains for product safety and shelf life.
(3) To describe the key principles and processes in temperature controlled supply chains.
(4) To outline possible future developments and constraints in this field.

Introduction

A quick examination of any full-range food superstore immediately brings home the importance of food to the consumer. The modern consumer expects the food in the store to be of good quality, to have a decent shelf life and to be fit for purpose. Similarly, if the retailer can present products attractively and extend their shelf life then there is more chance of the products being purchased and satisfying consumer needs. Managing the supply chain to maintain quality and 'fitness' therefore has direct cost and service implications. This process of distribution management, however, is not simply a question of moving box A from field B to store C. Many dimensions have to be managed, as other chapters in this book have identified. One of these dimensions is the need for a proper temperature regime: this is the focus of this chapter.

A food superstore contains products supplied and retailed at a number of different temperatures. Failing to maintain an appropriate temperature control can adversely affect the product's appearance or shelf life, at one end of the spectrum, or could potentially make consumers ill or even kill them at the other end. Temperature controlled supply chains could be said to be a matter of life or death.

This chapter focuses on the issues raised by temperature controlled supply chains (TCSCs). It begins by offering a definition and some thoughts on the importance of TCSCs. A discussion of the changing nature and implications of TCSCs follows this. Selected issues are then highlighted for further discussion. Finally, future directions and issues for TCSCs are presented.

What is a temperature controlled supply chain?

At its simplest, a TCSC is a food supply chain which requires food products to be maintained in a temperature controlled environment, rather than exposing them to whatever ambient temperatures prevail at the various stages of the supply chain. This basic description hides, however, a complex, potentially complicated and expensive process. The length and complexity of such a supply chain is determined by the natures and sources of the products, the legal and quality assurance requirements on food safety, and the distribution facilities available from production to consumption.

There are several temperature levels for food to suit different types of product groups. For example, we might identify frozen, cold chill, medium chill and exotic chill: frozen is –25°C for ice cream and –18°C for other foods and food ingredients; cold chill is 0°C to +1°C for fresh meat and poultry, most dairy and meat based provisions, most vegetables and some fruit; medium chill is +5°C for some pastry based products, butters, fats and cheeses; exotic chill is +10°C to +15°C for potatoes, eggs, exotic fruit and bananas.

If a food supply chain is dedicated to a narrow range of products, the temperature will be set at the level for that product group. However, if a food supply chain is handling a broad range of products then an optimum temperature or a limited number of different temperature settings are used. Failure to maintain appropriate temperature regimes throughout a product's life can shorten the life of that product (it goes 'off') or adversely affect its quality or fitness for consumption. The temperature regimes that are utilised after a consumer has purchased the product remain uncertain, but are not the main focus of this chapter.

From the description of the types of temperature needed and knowledge of the supply chains in food retailing from earlier chapters of this book, it should be immediately obvious that the management process in a TCSC is a complicated one. Chilling and freezing products is in itself difficult, but maintaining this throughout a product's life and in both storage and transit is complicated. How, for example, can a retailer ensure that products are always under the appropriate temperature regime when they travel from a field in New Zealand to a refrigerator in a house in Auchtermuchty? In addition to complexity, however, we can argue that the TCSC is also an increasingly important channel, both in absolute volume terms and due to risk assessments of chain failure.

Importance of temperature controlled supply chains

We have already implied that food safety and health are reasons enough for interest in TCSCs. We expand on this later, but also suggest further reasons why TCSCs are of importance.

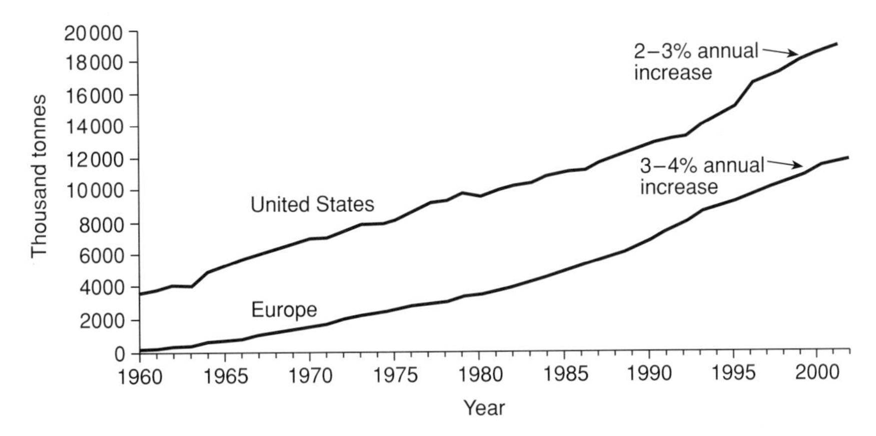

Figure 12.1 Frozen food consumption in the USA and Europe from 1960 to 2000 (Source: Quick Frozen Foods International, 2002).

The TCSC in food retailing accounts for a significant and steadily increasing proportion of the market (McKinnon & Campbell, 1998). CH Robinson/Iowa State University (2001) suggest that over half the spend in American supermarkets is on temperature controlled products; in the UK, frozen food has been increasing in volume by an average of 3–4% per annum for the last 40 years (Figure 12.1); developments in products such as ready meals and prepared salads have further expanded the market. Analysts predict that the meal solution sector will continue to increase very rapidly (Gorniak, 2002). 'Fast food' chains have captured a huge market share and are reliant on frozen product. The importance of products requiring temperature control, both to the consumer and to retailers, has thus been increasing, and seems set to develop further.

Even products that we take for granted require some form of temperature control. Sandwiches, for example, require chilled storage of ingredients which are then combined to make the finished product, that in turn requires temperature controlled storage, distribution and display (Smith *et al.*, 2001). Failure to maintain adequate control (e.g. placing prawn sandwiches in the sun) generates obvious risks. More subtly, an inability to maintain temperature control will reduce shelf life for the product, which is in any case often very limited; this increases wastage and complicates the supply dynamics, adding costs. Similarly, much of the bread in supermarket in-store bakeries is brought to the store frozen, and baked/heated on the premises.

This growth in temperature controlled products derives from changes both in consumer behaviour and preferences, and in the supply systems' abilities to respond to these changes. For example, consumer lifestyles and shopping patterns have altered dramatically, raising demands for convenience products. Production of ready meals has matched a desire for food heating rather than food preparation. The ability to take a frozen meal from the freezer and microwave it for immediate consumption meets the needs to save time and for convenience. Ready washed and prepared salads fulfill the same requirements. At the same time, greater concern over freshness of products, such as meat and fish, has focused attention on their handling in the supply chain.

On the supply side, changes in the location of product sources and the removal of wholesalers from the channel have had major effects. Technological changes in production and distribution have also allowed a transformation of the supply network. As production and distribution technological capabilities have developed, so the ability for national and international, rather than local, sourcing and distribution has emerged. Products can be brought across the world to satisfy demands for products 'out-of-season' or of an exotic nature, as well as for reasons of lower purchase or cost price. Internationalisation of supply of even indigenous products is common. The system developments needed to meet the demands for quality and consistency, including temperature control aspects, do impact on the channel composition (see, for example, Dolan & Humphrey, 2000). The handling systems to manage the air freighting of, for example, tomatoes from the Canary Islands, baby sweetcorn from Egypt or flowers from Malaysia require considerable technological development; they also represent a fundamental organisational and relationship shift.

It has been argued by CH Robinson/Iowa State University (2001, pp. 1–2) that TCSCs are more important than 'ordinary' retail supply chains because they have inherently more complexity and complications:

> 'The (Logistics) challenge is more formidable when the materials and products require temperature control. The shelf life is often short for such products, placing even greater importance on the speed and dependability of the transportation and handling systems. Temperature controlled products also require specialised transportation equipment and storage facilities and closer monitoring of product integrity while in the logistics system.
>
> Adding to the logistics complexity is the seasonal demand for many temperature controlled products . . . arising from natural production conditions and consumer demand . . . In addition, carriers of temperature controlled products confront unique requirements and incur greater costs than carriers of dry products.'

Some of the uniqueness and increased costs derive from this need to ensure temperature control. There is extra cost incurred in the requirements for handling temperature controlled products, and also in the need to monitor temperature regimes in the supply chain.

As the number and range of temperature controlled products have increased, and a number of market failures have occurred, so the issue of food safety has become more central (Henson & Caswell, 1999). Failures of food safety in the UK (not all, of course, associated with failures of temperature control) are common on a localised and individual level. For example, there is a high level of personal food poisoning in the UK, although the extent to which this is a result of product or channel failure rather than an individual consumer's lack of knowledge or care is unclear. More publicly notable, however, have been national events ('food scares') such as *Listeria* in cheese, *Salmonella* in eggs and chickens, BSE in cattle and *E. coli* 157 in meat. These national events raise concern and increase comment about food safety. There are thus perceptions concerning the safety of supply of food and food chains, which in turn have focused attention on risk assessment and risk management. Therefore TCSCs gain importance from the risks associated with their failure and from the steps necessary to minimise these risks, some of which are voluntary and company specific whilst others are required as the result of legal developments over the past decade.

As a consequence of risk assessments and the major problems in food safety, TCSCs have become a focal point for the development of food safety legislation across Europe which has introduced requirements covering a broad range of issues, one key aspect being the temperature conditions under which products are maintained. Such legislation, combined with increasing retailer liability for prosecution, has put great pressure on the standards of control throughout the food supply chain, particularly in the case of temperature control. For these reasons, TCSCs are often seen as a specialist discipline within logistics. To some extent this is understandable given the need for specialist facilities, e.g. warehouses, vehicles and refrigerators, to operate chilled or frozen distribution channels. This specialist market, however, is itself increasing in scale and scope, both as the market expands and as operational and managerial complexity increases.

It is not all cost and regulation, however, because there are operational and commercial benefits to be gained from proper TCSC management. These might include an increase in shelf life and freshness, and thus better customer perception of products and the retailer. This increase in product quality and perception is the direct result of maintaining the correct temperature for that product group steadily and constantly throughout its supply chain journey. One major effect of an increase in shelf life and freshness has been that consumers can notice the difference between product supplied through a fully TCSC and that supplied through a partially TCSC, and so make product and retailer choice decisions accordingly. Whilst it is generally the case today that the major food retailers maintain chill and cold chain integrity and thus have totally controlled TCSCs, this has not always been the case. TCSCs have thus changed considerably over the past two decades.

Changes in temperature controlled supply chains

The TCSC has developed and changed since the 1980s. In the UK, the supply chain formerly consisted of single temperature warehouses dedicated to narrow food product ranges, e.g. butters, fats and cheeses at +5°C, dairy based provisions, meat based provisions, fresh meat and poultry, fruit and vegetables, and frozen products. The design, equipment and disciplines were only partially implemented so that there was incomplete integrity of the temperature control. Products were exposed to periods of high ambient temperature, which affected their shelf life and quality. Single temperature systems also meant that many more deliveries were needed. Such systems were essentially inefficient and ineffective.

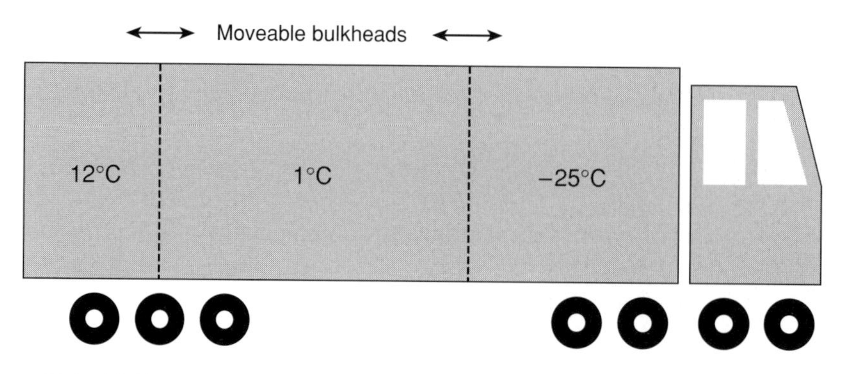

Figure 12.2 Multi-temperature trailer design.

Such a situation existed in the 1980s in Tesco (see Sparks, 1986; Smith & Sparks, 1993; Smith 1998) when the Tesco TCSC consisted of a large number (twenty-seven) of small single temperature warehouses, each specialising in the storage, handling and delivery of a narrow product range. Examples of these sets of product ranges were: fresh produce; fresh meat and poultry; butters, fats and cheeses; chilled dairy provisions; chilled meat provisions; frozen foods. Each set was managed by a different specialist logistics service provider organised on behalf of the manufacturer and supplier. The deliveries to the retail stores took place two or three times a week using the temperature controlled vehicle (see Figure 12.2) going from one store to another delivering the appropriate number of pallets of products. The delivery notes and product checking were conducted at the back door of the store and the cost of delivery was included in the price of the product. Fresh meat and poultry was controlled on an individual case basis and charged by weight as each case had a different weight (variable weight charging).

There are several limitations of this model of a TCSC. It was expensive to expand to meet large increases in overall growth in volume because it required the building of more and more single temperature warehouses. The retail delivery frequency was limited. The delivery volume drop size per store was small and vehicles used were 'under-sized' because of problems over retail access. Also at that time, the importance of maintaining total integrity of the chill chain was not fully appreciated.

The strategy decided upon by Tesco was to build a small number (seven) of new large multi-temperature 'composite' warehouses that would store, handle and deliver the full range of product sets, all from the same location. The composite temperature regimes are frozen at –25°C, cold chill at +1°C, and exotic chill at +12°C. The manufacturers and suppliers of all the product groups now made daily deliveries into the composite distribution centre (Figure 12.3 shows a Sainsbury modern composite distribution centre); the composite delivery frequency to the retail stores was increased to daily using delivery vehicles with movable bulkheads and three temperature controlled evaporators so that up to three different temperature

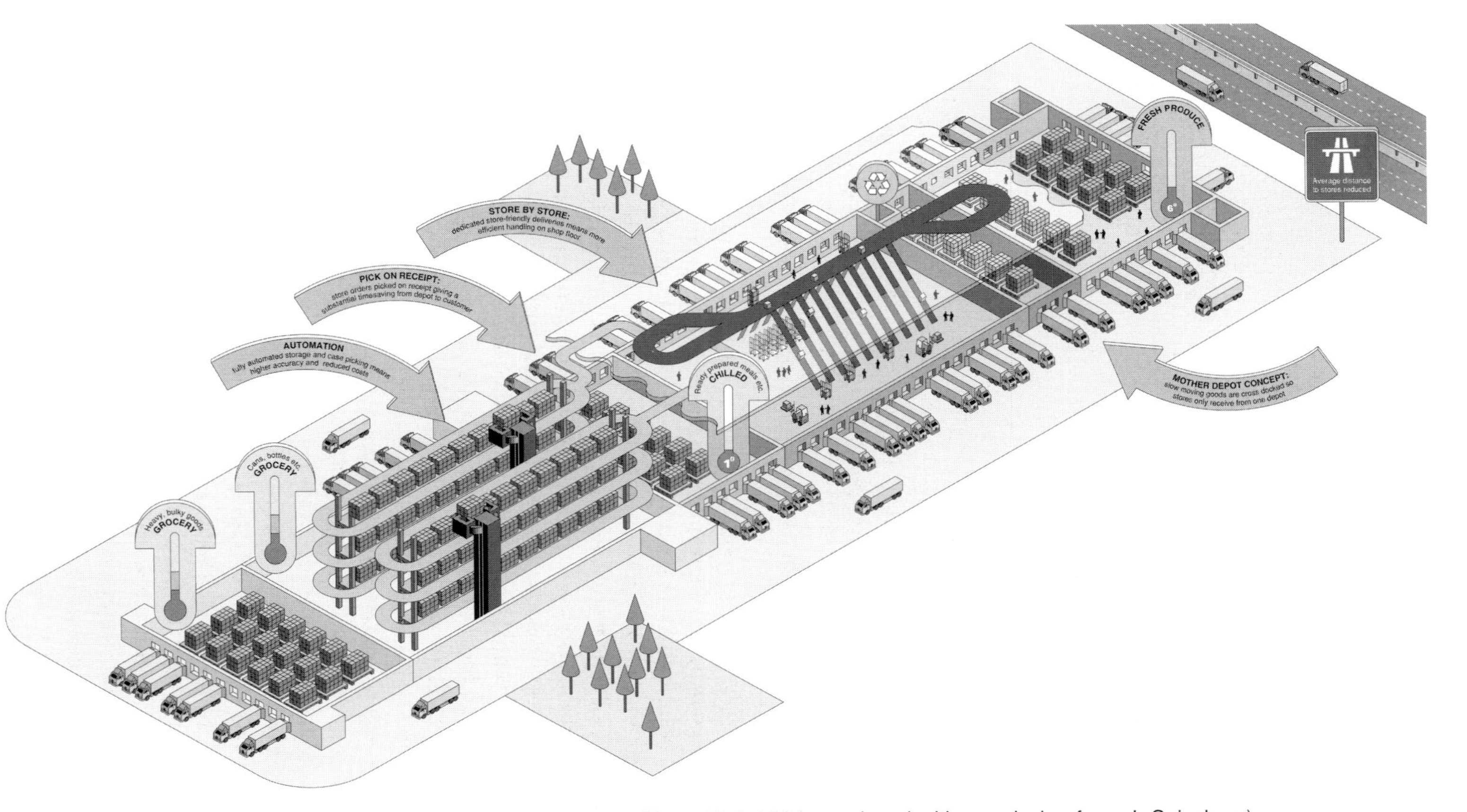

Figure 12.3 Sainsbury's modern composite distribution depot at Hams Hall, UK (reproduced with permission from J. Sainsbury).

Figure 12.4 Internal bulkheads for temperature separation in a multi-temperature trailer.

regimes could be set on the one vehicle (Figures 12.2 and 12.4). The benefits were improved vehicle utilisation and improved service to retail. Chill chain integrity disciplines were implemented rigorously from supplier to distribution centre (Figures 12.5, 12.6) to retail shelf.

There are other aspects to these changes. Distribution and retail agreed a policy of not checking the goods at the retail backdoor; this increased the speed with which the goods could be transferred into the temperature controlled chambers at the store and so improved chill chain integrity. The goods were delivered in green reusable plastic trays, on 'dollies', or on roll cages, which improved handling at store, both in terms of speed and quality. New store designs permitted the use of full-length vehicles, so improving efficiency.

Another major change in supply chains between the 1980s and 2000s has been the increasing pace of the order and replenishment cycle (McKinnon & Campbell, 1998). Today, with many fresh products there is no stock held in the retail distribution centre overnight (IGD, 2001). Stock holding in frozen products has also declined to below 10 days (IGD, 2001). Lead times have continued to be reduced. One of the key drivers of this increase in pace has been the development of information technology which has enabled a large volume of data to be collected, processed and transmitted at faster speeds. Today data are collected from the point of sale and used in calculating future customer demand, which in turn forms the basis of the orders placed on suppliers. The scale, control and skill of the retail logistics operation has improved, so that even distance sourced products can be rapidly

Figure 12.5 Open-air unloading of ambient product.

Figure 12.6 Temperature controlled dock at distribution centre.

transported to their destinations at the regional distribution centres. The move to centrally prepared meat and poultry rather than having butchers at the retail stores is one example of this. Another example is the sourcing of produce from Spain directly from the growers into the distribution centres (see Case Study 12.1). These changes, encouraged by information technology, amongst other factors, require modifications to supply chain facilities and operations to ensure chill chain integrity.

Case study 12.1: Produce direct from Spain

Spain has become one of the major providers of produce to the rest of Europe. In the late 1990s, major UK supermarkets started to purchase produce directly from Spanish suppliers rather than through UK wholesalers. The total direct flow of produce from Spain solely to the UK is over 1000 vehicles per week.

For example, Iceberg lettuce is grown in large volumes between October and May in Murcia and Almeria in south-east Spain under direct contract between the retailer and the growing co-operatives. The retailers' quality assurance and technical departments provide the grower with the product specification and transport temperature control requirements from Spain to the UK.

There are two methods of direct delivery into the UK supermarket distribution network. The first is to fill the vehicle in Spain solely with Iceberg lettuce. The delivery is split after it arrives in the UK by sending the vehicle to two distribution centres. The second method is to combine several produce products, e.g. Iceberg lettuce and courgettes, while the vehicle is still in its originating region in Spain. This combined product volume fills the vehicle, which then delivers the whole load to a single distribution centre in the UK.

The distance from Murcia to central England is 2400 km (1500 miles). The deliveries flow through daily. The total process from harvesting to customer is four days:

- *Day 1*: The Iceberg lettuce is harvested, cooled, packaged and loaded into temperature controlled vehicles set at +3°C.
- *Day 3*: The Iceberg lettuce arrives directly at the UK supermarket temperature controlled distribution centre where it is checked in. Within 3 hours it is allocated and picked for a retail store, ready to go out on the next delivery.
- *Day 4*: The Iceberg lettuce is on display in the retail store, available for the consumer to purchase.

The code life on direct Iceberg lettuce is one day above 'normal' deliveries. This extra day can be used for rolling stock in the distribution centre. The ability to roll stock means that full loads of Iceberg lettuce can be delivered directly to the distribution centre. Any stock that is not allocated to a store and picked immediately (due to demand) can be rolled over to supply the next day's orders.

The range of products delivered directly has increased from the original Golden Delicious and Granny Smith apples to include now Braeburn and Royal Gala apples on top fruit; white seedless grapes; nectarines; peaches; Iceberg lettuces; Galia melons; broccoli.

The seasons of other products obtained directly from Spain are as follows: December–January, soft citrus; January–May, tomatoes, broccoli; June–August, Galia and Honeydew melons.

Source: Author interviews.

Following the implementation of centralised distribution, the attention turned to the condition of TCSCs for the inbound product groups from the supplier and manufacturer into the regional distribution centres. Examination of the logistics of the inbound supply chain revealed that there were huge opportunities to improve transport efficiency. The increasing pace of the retail supply chain had resulted in most suppliers of temperature controlled product sets sending their vehicles long distances, but only partly filled, to the various retailers' regional distribution centres. So, for example, suppliers' vehicles carrying fruit and vegetables from a region such as Kent were following each other, partially filled, to the same distant regional distribution centres in northern England and Wales. Clearly, there were opportunities for the consolidation of supply.

This process of consolidation saw the appointment of designated logistics service providers in the appropriate regions to manage and operate temperature controlled consolidation centres, accumulating full vehicle loads of temperature controlled products to despatch to the composite distribution centres. These consolidation centres also conducted quality assurance testing of the product because two benefits accrued: the first was that the consolidation centres were close to the suppliers so that any problems could be dealt with face to face where required; the second was that the vehicles did not need to undergo quality assurance checking when they arrived at the distribution centre. This improved the turnaround time of the inbound vehicle, increasing its productivity and profitability, and also enabled the handling operation to commence earlier and so keep the goods-in bay clear for the next set of deliveries, this being especially important in the early evening when a very high volume of produce harvested that same day was delivered.

Some of the effects of these changes to the Tesco supply chain are considered in Table 12.1. This summarises the last 15–20 years of temperature controlled supply and the ways in which this has changed. Over the time-period the shelf life for these products has increased considerably. This provides better product for longer for the consumer, and is more efficient for the retailer. It does require a major reorientation of the supply chain and a dedication to standards. The overall effect, however, has been to provide fresher product more quickly and cheaply to the retail store and to lengthen the shelf life and quality time of a product for the consumer.

The discussion above is centred on developments in Tesco. Similar operations and developments have been introduced by other major food retailers. These have been needed to handle the massive expansion of demand in the temperature controlled sector in recent years and to compete with the market leader. Not all the changes and developments can be outlined above. Changes to packaging and transport environments, e.g. vacuum or other atmosphere controlled packaging, have also extended the shelf life of products and improved in-store quality.

Table 12.1 Tesco case study: enhancements in shelf life.

Stage	Shelf life (days)			Temperature controlled supply chain status and improvement action
	Soft fruit	Top fruit	Veg	
Pre 1980				Three single temperature produce centres. Ambient and +5°C. Code dates not a legal requirement. Shelf life managed at retail. Retail ordered from local suppliers without any technical support.
1980 to 1986				Two further produce centres. Operating procedures remain the same. Supplies normally loaded in yard or from ambient bays. Many vehicles have curtain sides.
1986				Notice given that code dates are to become a legal requirement for produce. Produce technical team establish shelf life and introduce quality checks at distribution centres. 78–48 hour ordering cycle to retail.
1987	2 CD + 2 CL = 4	5 CD + 2 CL = 7	3 CD + 2 CL = 5	Code dates (CD) introduced for loose and pre-packed produce. In addition to the selling CDs there were additional days where product would be at its best; this period was called 'customer life' (CL). Introduction of quality checks in produce depots to enforce specification. Two further produce centres. Code of practice introduced for suppliers includes distribution centre controls and vehicle standards, e.g. no curtain siders.
1989	2 CD + 4 CL = 6	5 CD + 6 CL = 11	3 CD + 5 CL = 8	Six composite distribution centres opened. Separate temperature chambers of +3°C, +10°C, +15°C for produce. Composite multi-temperature trailers deliver at +3 and +10°C, loading from sealed temperature controlled loading bays. Customer life extended by 2 days for soft fruit, by 4 days for top fruit and 3 days for vegetables. No increase in CDs. Consumer demand for fruit and vegetables doubled as a consequence of the introduction of strict temperature control disciplines throughout the supply chain.

1990				Food Safety Act: to meet due diligence, Hazard Analysis Critical Control Points (HACCP) introduced throughout the supply chain: the result was more consistent shelf life but no increase in days. Retail stores only allowed to buy from suppliers with technical approval.
1995	2 CD + 5 CL = 7	5 CD + 7 CL = 12	3 CD + 6 CL = 9	Produce temperature controlled consolidation hubs introduced. Six further hubs added over next three years. Quality assurance control introduced at hubs so quality checked before produce despatched to the composites. Hubs located close to supplier regions so prompt resolution of problems with supplier management. Shelf life review shows increase of one day across all vegetables, soft fruit and stone fruit. Salads become inconsistent because of harvesting during the night before dew point. But there was greater benefit of starting despatch earlier, especially from Spain. Retail order lead time 48–24 hours.
1997				Composite distribution centres change produce chamber temperatures to +1 and +12°C with tighter variation of ±1°C from ±2°C before. No change to shelf life because supply chain disciplines were fully in place.
1998	2 CD + 5 CL = 7	6 CD + 7 CL = 13	4 CD + 6 CL = 10	Technical departments given targets to increase produce shelf life. One potential improvement was to introduce USA type variety control. Benefit only possible because of very strict total supply chain temperature control. One extra day of CD for stores.
2000				Continuous replenishment introduced. Benefit is split deliveries into retail stores with different CDs for retail without any loss of CL.
2002	2 CD + 5 CL = 7	7 CD + 7 CL = 14	5 CD + 6 CL = 11	Further supply chain improvements in shelf life to extend CDs by one day; no change to CL to improve availability on selected lines, i.e. core vegetables, top fruit and stone fruit, but not salads or soft fruit. Three potential methods are: (a) atmospheric control, especially during the three-day delivery from Spain; (b) humidity control; (c) ethylene control.

Source: Author interviews.

Issues in temperature controlled supply chains

The discussion above and comments in the introduction allow the identification of a number of key issues in TCSCs. Here, three are identified for further discussion: the issues of costs, food safety and HACCP, and partnerships.

Costs

The basic supply configuration in the temperature controlled channel is not very different to those in ordinary retail distribution channels. The demands placed on the components, however, are far more extreme and thus the issue of costs of facilities and operations is important.

TCSCs place strict constraints on the design, equipment and discipline of the operation, which make the cost greater than for ambient products. Temperature controlled storage facilities need to be maintained at the appropriate temperature with accurate recording equipment and cooling equipment, and be able to cope with high ambient temperatures, especially in the summer. Vehicle docking bays need air bags that inflate around the vehicle to prevent exposure of the product to ambient temperatures. For frozen storage facilities, the loading and unloading bays should be at 0°C. Vehicles require appropriate insulation and refrigeration, with control to set and maintain the correct temperature. An important facet of this transport refrigeration is that it is not designed to remove heat from the product (as in 'normal refrigeration') so it is essential that the heat is taken out of the product before it is loaded onto a vehicle. If not, heat will transfer to other products causing them to be exposed to a temperature outside the designated range. Some vehicles have bulkheads and several evaporators so that different sections can be set at different temperatures: the benefit is that vehicle utilisation is improved, but operating procedures are made more complicated. This also affects costs. The cost of a multi-temperature refrigerated trailer is £100 000 compared to £30 000 for an ambient trailer. The cost of warehousing is £186 per sq. metre (£20 per sq. foot) compared to £93 per sq. metre (£10 per sq. foot) for ambient. Warehouse operatives and drivers must behave in accordance with the requirements for chill chain integrity to protect the product. The cost of losing a trailer load of product through overheating is not only expensive, but also severely impacts on service level to retail and the consumer because the pace of the supply chain does not leave time to recover with alternative product. Such cost considerations have enabled niche operators to enter and develop the market for frozen and chilled distribution. There is also a specialist association in the UK to assist this sector of the logistics industry (Cold Storage and Distribution Federation – http://www.csdf.org.uk/) and to liaise with government on regulations.

Food safety and HACCP

As noted earlier, the integrity of temperature controlled supply chains is important for food safety. This places an obligation of care and duty of implementation on the supplier, retailer and logistics industry. In the UK for example, the Food Safety Act of 1990 defined the storage, handling and transportation requirements for food products, including temperature control for certain categories. One of the requirements of this Act is that it makes it an absolute offence to sell food that is unfit for human consumption; clearly, food that has 'gone off' due to inadequate temperature control can fall into this category. The Act, however, allows for a defence of 'due diligence' against any charges. Thus a business may be able to mount a defence based on evidence that all reasonable precautions had been exercised to avoid committing the offence. With regard to temperature control, this implies that systems of control, maintenance, monitoring and recording (for evidence) of the temperature regimes in the supply chain are needed.

The UK Food Standards (Temperature Control) Regulations of 1995 made it an offence to allow food to be kept at temperatures that could cause risk to health, again requiring a tightening of systems in the chain. This was effectively codified by the General Hygiene Act of 1995, which required all food businesses to adopt a risk management tool such as Hazard Analysis Critical Control Points (HACCP). Loader & Hobbs (1999) see this as a change in philosophy, representing a move away from an end product food safety inspection approach to a preventative, scientific focus, with the responsibility for risk management placed on the food business proprietor. As a result, HACCP and other systems (Sterns *et al.*, 2000) have been vital in establishing process controls through the identification of critical points in the process that need to be monitored and controlled (see Case Study 12.2).

In the UK, these Acts were in essence national responses to approaches being recommended in Europe and codified in EU legislation. The UK food scares of the late 1990s also brought forward a response. The Food Standards Act 1999 created the Food Standards Agency (FSA) in April 2000 with the intention of inducing all those involved in the food supply chain to improve their food handling practices, including temperature control. There is no doubt that as the FSA becomes more established it will have a stronger role to play in TCSCs than has been seen to date (http://www.foodstandards.gov.uk).

Partnerships

This onus on due diligence and the responsibility of businesses had a major effect on systems of control and monitoring of performance. It also, however, had an effect on the business relationships and governance in place. If retailers, for example, wish to be protected from claims, in addition to their own practices, they have to ensure that their suppliers are undertaking good practices. This is true not only for retailer brand products, but for all sourced products. As such, traceability and tracking become more fundamental and good partnerships become crucial. Because costs rise when introducing new systems, increasing the depth and quality of partnerships is both a safeguard and offers possible cost benefits. As a result, in the UK partnerships expanded considerably after 1991 (Wilson, 1996; Loader & Hobbs, 1999; Fearne & Hughes, 2000). Food retailers today are keen to have such

Case study 12.2: Hazard Analysis Critical Control Points (HACCP)

It is important in the application of the disciplines of an integrated TCSC to understand the principles of the obligations of suppliers, retailers and logistics service providers. All have a duty of care for the product: to meet this duty of care they must demonstrate that they have applied due diligence in the structure and execution of their operation, i.e. that they have taken all reasonable methods to ensure the care of the product.

One of these reasonable methods is Hazard Analysis Critical Control Points (HACCP) and is central to the discipline of chill chain integrity in logistics. The quality assurance department conducts a survey of the supply chain under its control with the objective of identifying those circumstances where the product might be exposed to unsuitable conditions, i.e. hazards. It ranks these hazards according to the importance of their risk, e.g. high, medium, low. Procedures are then put in place at an appropriate level to prevent that risk. So, to express this differently, identify the hazards, analyse their importance, identify which are critical and set up control procedures at these points. Once HACCP has been put in place it becomes a strong argument that due diligence is being practised.

For temperature controlled supply chains, there are big benefits from putting the physical and operational procedures in place along the whole length of the supply chain. This investment reduces a high risk to a low risk. By stabilising the temperature throughout the life of the product, suppliers and retailers can concentrate on other aspects which can add value to the product, e.g. growing different varieties.

If we take as an example the movement of chill goods from distribution centre to retail stores on multi-temperature vehicles, the risk to food safety is high and the risk of occurrence is high. The critical control points for loading at the distribution centre are the following:

- Temperature setting stated on load sheet and run sheet.
- Loader checks load sheet and sets temperatures for compartment; loader secures bulkhead.
- Loader switches refrigeration on and ticks relevant temperature on load sheet; when loading complete a supervisor checks settings and switches against load sheet and signs off if correct.
- Load sheet handed into goods-out office.
- Driver checks digital readings (usually at the front of unit, visible in rear view mirror) against load sheet and, if correct, signs off and hands it in to goods-out office.
- Goods-out clerk checks if temperatures on load sheet and run sheet match, and, if correct, allows vehicle to leave.
- Goods-out supervisor undertakes daily checks to ensure compliance.

Source: Author interviews.

partnerships and to use them in their marketing, as seen in the numerous 'farm assured' type schemes. Such partnerships and changes in organisation of the supply chain are not restricted to UK suppliers. Dolan & Humphrey (2000) show how in Africa the requirements of the leading UK retailers have transformed the horticultural sector in scale and operational terms, leaving smaller producers in a precarious position. This scale dimension is linked closely to the legal requirements and the costs of compliance and potential chain failure.

Future developments and constraints

TCSCs have undergone considerable changes in recent decades. This process is likely to continue, driven as it is by tightening legislation and risk awareness, the increased costs of supply, and the demands on the chain from increased volume and pace of operation. A number of future issues can be explored.

First is the question of risk and integrity. There are a number of 'gaps' in the current TCSC. For example, at the retail outlet, very few stores have a chilled reception area that docks with the incoming distribution vehicle; the majority have an ambient delivery bay that is exposed to the outside temperature. The delivery reception area requires an operational discipline such that chilled and frozen product is not left exposed to ambient temperature for more than 20 minutes. The retail operational staff have to move the chilled and frozen pallets, roll cages, etc. promptly into the relevant temperature controlled chambers. Finally, when the product is being taken to the chilled or frozen retail shelf or cabinet within the store for replenishment, the same 20-minute rule applies. The potential for problems is clear. Another 'gap' occurs, of course, from the time that a consumer selects a product in the store. The length of time from selection to purchase and transport home could be considerable and affect the product adversely (e.g. ice cream in summer). As electronic commerce expands, so issues of home delivery will confront much the same problem. If the homeowner has always to be present such services will be more limited and/or costly, but dropped deliveries of temperature controlled products increase the risks for the consumer.

Secondly, there is likely to be further technological development. In the future, electronic temperature tagging could become the norm, so that operators throughout the TCSC will be able to monitor current and previous conditions. Such monitoring could be real-time on-line for some products or could be packaging based for others, e.g. colour coded packaging changing colour if temperatures go outside allowable ranges. Such monitoring advances may also be accompanied by advances in packaging environments leading to enhanced shelf life.

Finally, there are issues brought about by present globalisation and partnership trends. Whilst there have been benefits to the introduction of current procedures and practices, concern is mounting about the environmental costs of monocultures and the extreme distribution distances that are travelled. Given the costs of compliance to meet western food safety concerns, it might be more beneficial to look for more local sourcing. The internationalisation of retailing acts as a counter to this,

Case study 12.3: Tesco Thailand

Tesco Thailand was founded in 1997 through the purchase of 13 hypermarkets from Lotus. By 2002 there were 35 hypermarkets comprising 0.41 million sq. metres (3.8 million sq. feet). 15 new stores are planned up to 2005, an annual growth rate of 36%. As the country has an average temperature of 25 to 30°C, it provides a contrast in temperature controlled supply chains with the UK.

Malai Chuangkrut, Quality Assurance Manager for Tesco Thailand Distribution, visited Tesco UK in July 2002 to study standards and techniques. She found two similarities and three major differences.

The similarities arise because Tesco Thailand implemented Best Practice from UK Tesco Secondary Distribution. This involved in particular, first, the building of a single multi-temperature distribution centre and second, the use of temperature controlled vehicles from distribution centre to retail store.

The first major difference is in scale. There is one distribution centre in Thailand, but many in the UK. The average size of suppliers in Thailand is small, both absolutely and in comparison with the UK. The second major difference is a high ambient temperature and its potential effects on produce. The third and most significant difference is the requirement to educate suppliers in the disciplines of Quality Assurance Best Practice as developed in the UK. Malai's role is to put in place a development programme for the suppliers. This programme covers HACCP analysis of the temperature controlled supply chain, imposing the disciplines to achieve temperature integrity in the primary transport. The aim is to develop a supply chain with these suppliers which is fully temperature controlled, and which will provide the commercial benefits for suppliers and retailers which have been demonstrated in the UK.

Source: Author interviews.

however, and may allow retailers to build deeper partnerships across the globe and to utilise their experience to enhance the quality of the local supply chain (see Case Study 12.3). The future organisational shape and the role of partnerships remain subject to change (Zuurbier, 1999).

Summary

This chapter has examined temperature controlled supply chains, focused on retail systems. It has argued that these are more complex and increasingly important to retailers and consumers. The risks associated with supply failure in this market have forced the introduction of legislation at national and pan-national levels. This in turn has encouraged the reconfiguration of TCSCs across the globe, aided by technological developments in facilities and operations. TCSCs are vital to businesses and the health and well-being of consumers. As such they will be subject to more critical appraisal and control in the future.

Study questions

1. Examine your weekly purchase of food products. Which of these require temperature control in their supply? Does this influence your product and store choice decisions?
2. Define TCSC and chill chain integrity.
3. In what ways are food safety issues and TCSC issues related?
4. Think about your handling of temperature controlled products after you have purchased them. Do you feel that you have appropriate procedures and standards and have identified all the risks from point of purchase to point of consumption?

References

CH Robinson Worldwide/Iowa State University (2001) *Temperature Controlled Logistics Report 2001–2002*. CH Robinson Worldwide, Iowa State University, Eden Prairie, IA.

Dolan, C. & Humphrey, J. (2000) Governance and trade in fresh vegetables: the impact of UK supermarkets on the African horticultural industry. *Journal of Development Studies* **37**(2), 147–176.

Fearne, A. & Hughes, D. (2000) Success factors in the fresh produce supply chains. *British Food Journal* **102**, 760–772.

Food Safety Act (1990) The Stationery Office, London.

Food Standards Act (1999) The Stationery Office, London.

Food Standards (Temperature Control) Regulations (1995) The Stationery Office, London.

General Hygiene Act (1995) The Stationery Office, London.

Gorniak, C. (2002) *The Meal Solutions Outlook to 2007*. Reuters Business Insight, London.

Henson, S. & Caswell, J. (1999) Food safety regulation: an overview of contemporary issues. *Food Policy* **24**, 589–603.

IGD (2001) *Retail Logistics 2001*. Institute of Grocery Distribution, Watford.

Loader, R. & Hobbs, J.E. (1999) Strategic response to food safety legislation. *Food Policy* **24**, 685–706.

McKinnon, A.C. & Campbell, J. (1998) *Quick Response in the Frozen Food Supply Chain*. Christian Salvesen Logistics Research Paper, Heriot-Watt University, Edinburgh (http://www.som.hw.ac.uk/logistics/salvesen.html).

Smith, D.L.G. (1998) Logistics in Tesco. *Logistics and Retail Management* (Fernie, J. & Sparks, L. eds), pp. 154–183. Kogan Page, London.

Smith, D.L.G. & Sparks, L. (1993) The transformation of physical distribution in retailing: the example of Tesco plc. *International Review of Retail, Distribution and Consumer Research* **3**, 35–64.

Smith, J.L., Davies, G.J. & Bent, A.J. (2001) Retail fast foods: overview of safe sandwich manufacture. *Journal of The Royal Society for the Promotion of Health* **121**(4), 220–223.

Sparks, L. (1986) The changing structure of distribution in retail companies. *Transactions of the Institute of British Geographers* **11**, 147–154.

Sterns, P.A., Codron, J.-M. & Reardon, T. (2000) *Quality and Quality Assurance in the Fresh Produce Sector: A Case Study of European Retailers*. Paper presented at American

Agricultural Economics Association Annual Meeting, Chicago, August 2000 (http://agecon.lib.umn.edu/).

Wilson, N. (1996) The supply chains of perishable products in northern Europe. *British Food Journal* **98**(6), 9–15.

Zuurbier, P.J.P. (1999) Supply chain management in the fresh produce industry: a mile to go? *Journal of Food Distribution Research* **30**(1), 20–30.

Chapter 13

Factors Influencing Supply and Demand for Organic Foods

G. Butler, H. Newton, M. Bourlakis & C. Leifert

Objectives

(1) To give an overview of organic farming legislation/standards.
(2) To analyse the current situation in the organic food market.
(3) To discuss the factors that affect demand and supply in the organic market.
(4) To provide models that can be used to predict future development of the organic food market.

Introduction

Organic retail market sales have increased exponentially throughout central and northern Europe in the last 10 years and now account for 2.9% of food sales in the European Union (EU) (Soil Association, 2000; Anon., 2000; Lampkin & Measures, 2001; UKROFS, 2001).

Over the past five years, the area of land used for organic production has more than quadrupled resulting in more than 3% of utilisable agricultural area (UAA) now being managed organically. However, the more rapid expansion of supply compared to demand has resulted in a narrowing of the gap between demand and supply (Lampkin & Measures, 2001; Hamm *et al.*, 2002).

This review will give an overview of organic farming legislation/standards, the current situation in the organic food market, the factors that affect demand and supply, and the models used to predict future development of this market.

Organic farming: definition and legislation

Organic farming is based around principles of producing food without chemo-synthetic fertility and crop protection inputs, and with minimal use of veterinary inputs and food additives (Lampkin, 1999). A broad outline can be seen in Table 13.1.

Table 13.1 Important characteristics of organic vs. conventional farming systems (Lampkin, 1999).

	Organic farming	Conventional farming
Crop production	• Long rotations (6–9 yrs) • No synthetic pesticides — S and (Cu) based fungicides — biological pest/disease control — plant extract based products • No water soluble N&P fertilisers (no KCl) — K_2SO_4 lime and microelements if necessary (soil analysis)	• Short rotations (1–5 yrs) • Synthetic pesticides used — 220 herbicides — 186 pesticides — 143 fungicides • NO_3, NH_4, urea, KCl, superphosphate, main NPK inputs
Animal production	Focus on welfare and on farm sustainability • Access to outside • Lower stocking density • No growth promoters • Longer withdrawal periods • On farm feed production	Focus on maximum production efficiency • Indoor production • High stocking densities • ABGPs use (pigs/poultry) • Standard withdrawal period • High proportion of bought in feed
Food processing	< 40 additives allowed • Hypochlorite use (as sanitising agent) prohibited	300 additives allowed • Hypochlorite use (as sanitising agent) permitted (< 200 ppm)

The legislation governing organic food production via the EU is based upon European Commission (EC) regulations 2092/91 and 8697/98 (EC, 1991, 1999) that form the baseline with which all member states and sector body organic standards in the EU have to conform. Whilst other organic standards prescribed by individual member countries and/or certification companies may be more 'restrictive' than these, they may not omit any of the standards laid out in the EU directives. Each EU country and EU based organic sector body can either use the EU standards or develop their own more 'restrictive' standards.

The auditing of organic farms for compliance with organic standards is carried out by certification agencies/sector bodies. In legal terms these sector bodies may be charities, government bodies or companies, and a sector body may include restrictions in addition to those set by the EU and member state legislation.

The organic food market

The organic food market in Europe is subject to large fluctuations both in supply and demand (Foster, 2000; Hamm *et al.*, 2002). The current market situation and the factors affecting demand and supply are discussed below to provide an understanding of the dynamics of the marketplace. It should be stressed that the relative importance of individual factors may differ between European countries and may change over time.

Demand

Recent reports into the organic food market found that, of the European countries studied in great detail, few had exactly the same drivers for organic market growth (Anon., 2001; Hamm *et al.*, 2002). Demand in all countries was driven by issues associated with general health and well-being. The markets in the UK, France and Germany were primarily driven by demand resulting from consumer fears about BSE and other food scares, in Sweden by consumer concern for animal welfare and the environment, and in Italy fear about GMOs was the most important driver (Anon., 2000; Soil Association, 2000; Lampkin & Measures, 2001). Consumers around Europe consider that the main barriers to purchasing more organic food are high price, limited range, no guaranteed availability, unsupported health claims and no significant improvement in taste, although all these barriers are thought to be decreasing in importance (Anon., 2000, 2001).

Price

The price premium of organic food (Table 13.2) is currently and expected to remain the greatest barrier to future expansion of demand (Anon., 2000). When data on price premiums (Table 13.2) are compared with data on consumer participation in different commodity groups (Table 13.3) a correlation between price premium and consumer participation can be found in all European countries with high levels of consumer participation in organic food.

More recently, however (Soil Association, 1998; EC, 1999; Anon., 2000, 2001), price premiums have fallen for many commodities because of increases in efficiency in primary production and reduced wastage levels in the supply chain (processing and retailing). Economies of scale in organic production, changes in the Common Agricultural Policy/subsidy system, and additional costs in conventional farming associated with the loss of agrochemical input (especially in the horticultural sector) are expected to further reduce the price differential between organic and conventional foods (Anon., 2000).

Quality

The quality of organic food available to consumers is a major barrier to organic market growth (Anon., 2001). Consumers demand equivalent or better sensory quality for organic foods compared to conventionally produced food. However, for certain vegetable and fruit commodities, lower quality produce is being presented for sale. This can be due to the shelf life of organic products being shorter (often due to problems in achieving the same levels of pest and disease control as in conventional farming) and/or due to demand still exceeding supply of organic food

Table 13.2 Consumer price premiums for organic products in 2000[a] (Hamm *et al.*, 2002).

	European Union						EFTA[b]	
	Aus	DK	Ger	UK	Ita	**EU**	CH	Nor
Organic meat								
Rump steak	**87**	5	29	**75**	86	40	2	48
Minced beef	23	47	47	**61**	N/A	45	23	32
Lamb chops	**96**	N/A	60	**78**	14	59	113	–7
Pork cutlet	50	27	**65**	14	13	51	80	74
Minced pork	**68**	49	**48**	N/A	92	52	20	N/A
Whole chicken	**142**	**267**	111	102	**191**	113	50	N/A
Milk products and eggs								
Milk	27	18	**56**	**59**	31	39	21	65
Butter	15	20	**72**	37	**77**	48	36	191
Natural yogurt	46	19	**176**	8	15	73	61	62
Cheese	15	23	**111**	43	47	58	41	27
Eggs	23	47	53	36	50	48	80	40
Vegetables and fruit								
Potatoes	34	13	**143**	100	54	91	104	39
Carrots	**82**	38	30	38	29	45	93	84
Onions	83	**128**	59	51	**122**	82	119	111
Tomatoes	**137**	**74**	**123**	88	45	89	113	81
Cucumber	67	71	**88**	N/A	1	70	79	82
Apples	49	56	57	63	29	45	46	102
Oranges	39	65	**125**	58	39	65	44	128
Wheat bread	62	47	33	60	**98**	61	800	140

[a] Figures in bold are more than 20% higher than the EU mean.
[b] European Free Trade Association.
N/A, not available.

Table 13.3 Organic food consumption by volume as a share (%) of total food consumption (Hamm *et al.*, 2002).

	European Union						EFTA[a]	
	Aus	DK	Ger	UK	Ita	EU	CH	Nor
Organic meat								
Beef	2.3	2.4	2.3	0.2	0.0	0.7	1.4	N/A
Lamb and goat	3.0	2.3	1.5	0.2	N/A	0.5	0.8	N/A
Pork	0.3	0.9	0.4	0.2	0.0	0.2	0.6	N/A
Poultry	0.3	0.4	0.3	0.1	0.1	0.1	0.2	N/A
Milk products and eggs								
Milk	6.4	10.6	0.9	0.8	0.4	1.0	4.2	N/A
Eggs	2.2	8.1	1.3	1.9	0.4	1.2	2.5	N/A
Vegetables and fruit								
Potatoes	3.6	7.4	1.5	0.7	0.2	1.3	2.2	N/A
Vegetables	3.4	7.3	2.1	2.9	0.2	1.1	3.1	N/A
Fruit (incl. nuts)	2.7	2.0	1.2	1.2	1.5	1.1	1.9	N/A

[a] European Free Trade Association.
N/A, not available.

for many commodities. As the market increases and supply gets closer to matching demand, quality is expected to increase (Anon., 2001).

One of the benefits frequently associated with organic food is that of improved taste (Soil Association, 1998). Studies have shown a link between organic production methods and improved taste in some commodities (Reganold *et al.*, 2001). However, the industry opinion survey (Anon., 2001) suggests that the lack of significantly better taste of organic commodities may be preventing a further increase in demand.

Availability

Both the range of products available and the continuity of supply affect the demand for organic produce. Consumers appreciate continuous supplies of commodities, regardless of season (Anon., 2001). As organic production increases, the range and availability of produce is expected to improve; recent predictions (Anon., 2001) suggest that by 2005 limited range is expected to become the least important consumer barrier to buying organic food. Limited availability and non-continuity of supply are also expected to become less of a barrier for demand.

State of the economy

In times of recession or difficult economic situations (e.g. high unemployment), growth in organic demand and the expansion of organic production area were shown to slow in parallel with consumers reducing spending on 'luxury items'. Whilst the effect is felt on all spending, it would be felt particularly on higher inherent value goods, e.g. red meat. The recent economic downturns in the UK and Germany have reduced land conversion rates to organic production (Soil Association, 1998, 2001; Anon., 2000; Lampkin & Measures, 2001).

Credibility of organic standards and auditing systems

The credibility of the organic industry and consumers' perception of the standards, sector bodies and certification agencies in general affect the demand for organic produce. Recent reports in UK newspapers claiming that organic pork from pigs weaned in a conventional system and then raised in organic finishing units was 'not really organic' (even though it complied with EU legislation) caused sales of organic pork meat to decrease by 40% and also a decrease in conventional pork sales, independently of whether it was produced in Germany or the UK (personal communication).

The credibility of the organic standards laid down by both EU agencies and UK agencies is important in gaining customer confidence in the organic industry. The establishment of the EU organic crop production standards in the early 1990s resulted in a boost in consumer confidence in organic standards and an acceleration of demand. However, loopholes in organic regulations, when identified by consumers, reduce customer confidence and could become detrimental to the development of the organic market (Hamm *et al.*, 2002).

Current disagreement within the UK and German organic sector bodies about standards and a diversity of different organic labels are thought to create confusion among consumers and compromise the positive image of the organic sector (Hamm *et al.*, 2000). Whilst Datamonitor (Anon., 2001) do not consider the image of the certification companies as a whole to have a large impact on the growth of the organic market, it is widely recognised that many of the negative images of conventional farming techniques and practices have spurred the growth of organics. Negative images of the organic industry may therefore force consumers to rethink their buying strategies.

Supply

Supplies of organic food are driven by producers, processors and retailers, but more recently also by national organic action plans focused on expanding supply of organic food (Anon., 2001). All these groups impact on the availability of organic food and therefore the growth of the marketplace. Without the retailers' drive to stock greater quantities of wide ranges of organic produce and processors diversifying to encompass a range of organic products, the industry would have grown much more slowly over the past five years in many EU countries (Hamm *et al.*, 2002).

Guaranteeing organic supplies is seen to be the largest obstacle to the future growth of the organic market. With demand currently exceeding supply for many commodities, retailers are finding sourcing products increasingly difficult, particularly with respect to continuity of supply (Anon., 2001). However, it is expected that as more land is converted, supply issues should become less critical.

Conversion rate

The main factor influencing organic food supplies is thought to be the conversion rate. Farmers converting their land to organic production have to wait two years, during which they have to manage their land to organic standards, before they can sell crops produced on the land as organic (for some livestock products, such as beef, the first organic products cannot be marketed until four years after the start of conversion). The conversion rate is therefore closely correlated to producer confidence in the organic market in two (for most crops and livestock products) to four years (for livestock products, such as beef) after the decision to convert has been made (see Table 13.4 for current producer and consumer price premiums; EC, 1991, 1999; Lampkin, 1999). Unless conventional producers can be confident in their ability in 2–4 years' time to produce organic food at a higher profit margin, they are less likely to take the risk of converting to organic production.

The minimum two year conversion period therefore remains the main obstacle to the expansion of organic supply. This is partially due to the difficulty in predicting future market demand, but also to financial losses during the conversion period

Table 13.4 Farmer price premiums (FPP) and consumer price premiums (CPP) for organic products in 2000[a].

	European Union						EFTA	
	Aus	DK	Ger	UK	Ita	EU	CH	Nor
Organic meat								
Beef (FPP)	17	12	33	173	25	34	10	7
Rump steak (CPP)	**87**	5	29	75	**86**	**40**	2	**48**
Minced beef (CPP)	**23**	**47**	**47**	61	N/A	**45**	**23**	**32**
Lamb (FPP)	20	25	35	43	N/A	43	43	12
Lamb chops (CPP)	**96**	N/A	**60**	**78**	14	**59**	113	–7
Pork (FPP)	50	66	71	120	40	69	80	74
Pork cutlet (CPP)	50	27	65	14	13	51	80	74
Minced pork (CPP)	**68**	49	48	N/A	92	52	20	N/A
Poultry (FPP)	200	370	200	170	100	182	108	N/A
Whole chicken (CPP)	142	267	111	102	**191**	113	50	N/A
Milk products and eggs								
Milk (FPP)	18	19	10	74	25	22	20	9
Milk (CPP)	**27**	18	**56**	59	**31**	**39**	21	65
Eggs (FPP)	100	250	177	154	100	167	100	N/A
Eggs (CPP)	23	47	53	36	50	48	80	40
Vegetables and fruit								
Potatoes (FPP)	280	84	300	296	N/A	257	132	109

[a] Figures in bold indicate commodities where the consumer price premium (CPP) is higher than the farmer price premium (FPP).
N/A, not available.
Source: percentage data from Hamm *et al.* (2002).

when yields and stocking rates are lower (because of the inability to use conventional inputs), but farmers are unable to sell products at organic premium prices (Lampkin, 1999; Lampkin & Measures, 2001).

Over the last two years the number of producers registering new land for conversion in the UK and many other EU countries has decreased (UKROFS, Soil Association, personal communication). Whilst most of the reduction is due to insecurity and uncertainty amongst livestock farmers (e.g. caused by the recent foot and mouth outbreak in the UK), there is also a more general underlying trend of a slowing down in the percentage annual increase in organic land area in many EU countries (Soil Association, 1998, 2001; Anon., 2000; Hamm *et al.*, 2002).

Data collected by Hamm *et al.* (2002) suggest that in the year 2000 the consumer price premiums for commodities (e.g. milk in Austria, Germany and Italy, lamb in Austria, Germany and the UK) where supply was similar or higher than demand exceeded the premiums obtained by producers. This may have affected producer confidence in the market and at least partially explains the decrease in annual percentage growth rate of organic land in many EU countries observed in 2001 and 2002 (UKROFS, Soil Association, Naturland eV, personal communication).

However, for commodities, such as poultry and eggs, where demand is much higher than supply, producers receive higher premiums than those experienced by the consumer (Table 13.4). Apart from an undersupply of increasing prices, this may be due to the processors/food retailers working in these sectors having introduced lower margins to increase turnover and achieve efficiencies of scale (especially with respect to waste reduction).

For conversion rates to increase in the EU, producers must have confidence in the gross margins being substantially higher than in conventional production in 2–4 years' time (by the end of the conversion period). While many sources suggest that price premiums are likely to decrease over time, this decrease is described as still leaving organic price premiums high enough to compensate for lower productivity (Anon., 2001).

Subsidy level

The level of subsidies that farmers receive to enhance their incomes over that generated by the sale of their produce can affect their decision making concerning conversion to organic production and ultimately the supply of organic commodities. The subsidies are divided into two groups, specific organic subsidies and general conventional subsidies.

If the level of conversion based subsidies increases, growers are increasingly likely to convert to organic. The subsidies enhance their income during the conversion period where production is reduced due to lower inputs and the price premium for organically grown food is not available because the land has not been in organic production for the set length of time and therefore is not yet deemed organic. For example, in countries such as Denmark, government conversion support is thought to have been a major driver for high conversion rates to organic production (Hamm *et al.*, 2002).

For conventional area or livestock-head based subsidies the situation is reversed. In some agricultural production systems where a large proportion of gross margin is made up of subsidies (see Table 13.5), the level of conversion to organic production has been low (Lampkin & Measures, 2001); this is particularly true for specialist arable crop producers. Although organic arable crops are eligible for arable area payments, the potential of increasing gross margins by conversion to organic is lower than, for example, dairy farmers, where only a small percentage of gross margin is made up of subsidies (Table 13.5).

In contrast, for beef and lamb production systems (especially extensive hill farming systems) the ability to supplement decreasing head-based subsidies (see Table 13.5) with organic conversion payments and the relatively low cost of conversion of such systems are thought to have resulted in the very rapid expansion of organic red meat production in countries such as the UK (Soil Association, 2001; Hamm *et al.*, 2002).

Table 13.5 Gross margins (GM; £/farm ha) and proportion of subsidies in the gross margin (%S in GM) for UK organic farming systems.

Farm type		Year		
		1996	1999	2001
Mainly arable	GM (£/farm ha)	707	753	736
	%S in GM	33	28	26
Stockless arable	GM (£/farm ha)	708	724	645
	%S in GM	47	37	33
Mainly dairy	GM (£/farm ha)	1331	1602	1550
	%S in GM	8	6	6
Specialist dairy	GM (£/farm ha)	1432	1700	1716
	%S in GM	0	0	0
Lowland livestock	GM (£/farm ha)	670	753	771
	%S in GM	29	23	21
Upland livestock	GM (£/farm ha)	271	273	246
	%S in GM	57	54	37

Source: Raw data used for calculations are from Lampkin & Measures (2001).

Organic standards

Changes to organic standards at the EU, national government or sector body level can impact on the availability and cost of organic food. An example of the potential impact of standard changes is the prohibition of the use of copper based fungicides for the control of late blight of potatoes. The derogation for the use of copper fungicides ended in 2002; with current varieties and cultural control methods the yield of organic potatoes is expected to reduce from the current 60% of conventional yield to 40% of conventional yield. This would increase production costs and/or the land area required for potato production to meet current demand. However, with premiums for organic potatoes already being above 100% (see Table 13.4), this would be expected to result in a decrease in consumer demand and could force organic potato production below the economic threshold.

The end of the derogation on the use of non-organic seed in December 2003 may also impact the supply of organic produce. The need to use organically produced seed may:

(1) limit production in EU countries which have not developed organic seed production systems or, more likely,
(2) force the EU to extend the derogation on the use of conventional seed which could result in reduced consumer confidence in organic foods.

A similar situation may occur in the livestock sector, where a shortage of certain organic feeds has resulted in a derogation to include small proportions of non-organic feed. This derogation is due to be removed in 2005, but low conversion rates of arable land may result in the need to extend derogations on organic feed, thus reducing the credibility of organic meat, milk and egg production systems with the consumer.

Supply chain and services infrastructure

The development of a successful organic production industry relies ultimately on the presence of a reliable infrastructure of supply industries, processors and advisors. Commercial producers have repeatedly expressed their need to be reassured that marketing channels, processors and technical information are available to them before making a decision on converting to organic production. Currently there are limited numbers of processors and marketing channels available for some commodities (e.g. oilseed rape, sugar beet). If these infrastructural problems are addressed, farmers will be able to produce and market a wider range of crops, which may improve the efficiencies of the entire crop rotation (for example, new 'break crops' can be introduced which reduce weed, pest, disease and fertility management problems in the rotations) (Robson *et al.*, 2002).

In areas such as dairy production, better infrastructures have enabled the industry to grow and meet the demand for organic dairy products. Co-operation between farmers, agricultural supply and services providers, processors and retailers is thought to be essential to prevent the bottlenecks and temporal under- and oversupplies for specific commodities (as currently experienced in some countries for lamb and dairy products) resulting in some organically produced foods having to be marketed through the conventional supply chains (Hamm *et al.*, 2002).

The presence of supermarkets as a major retailing force regarding organic products is thought to have improved supply chain infrastructures throughout the EU. Some supermarkets have set up organic groups to both source organic products and enable producers to effectively market them (Foster, 2000).

Infrastructures for technical and marketing advice and research are also less developed than those available to conventional farmers. The potential impact of optimising agronomic methods was demonstrated by Reganold *et al.* (2001) who developed organic apples production systems which have similar yields and production costs and higher product quality compared to conventional production systems in the US. With the development of 'organic action plans' in many EU countries and at the EU level, this is likely to improve in the short- to medium-term future.

Market prediction models

The development of the European organic market has been discussed in many recent publications. However, there is a lack of reliable statistics and both assessments of current market size and prediction models used vary greatly.

The European market for organic foods in 2000 was valued at US$9.6 million (Soil Association, 2000). It has been predicted that by 2005 the value may be nearer

US$21.4 million, a growth of 124% over five years across Europe (at an annual rate of between 16% and 19%) (Soil Association, 2000, 2001; Anon., 2001).

The prediction of consistent growth rates for organic production and organic retail growth enables a simple model to be produced for predicting future growth, and most predictions are based on simple stable growth rate assumptions. For example, Datamonitor consider that a 19% annual growth rate across Europe is likely over the next five years (Anon., 2001) whereas the Soil Association (2001) considers the annual growth rate to be around 25%.

By analysing data collected from European countries on participation in organic food purchasing and organic retail market growth (Anon., 2001b), it can be shown that the higher the percentage participation in the market (per capita consumption of organic food) in individual European countries, the more slowly the market (percentage annual increase in retail sales) grows. This correlation has been found to be significant and enables the production of a diminishing growth rate model (DGRM) for the expansion of the organic food market (C. Leifert & M. Bourlakis, unpublished observations). However, this model is only valid for the current set of socio-economic and consumer perception factors known to affect organic demand. If, for example, the price differential between organic and conventional foods decreases, if the quality and availability of organic products increases, and/or if the confidence in conventional foods decreases due to a new food scare associated with conventional farming practices, the market for organic food would increase. However, if the price differential between organic and conventional foods increases, if new organic standards are introduced which decrease production efficiency or quality of products, and/or if confidence in the organic standards or auditing system decreases, the market expansion is likely to be slower than predicted. The predictions resulting from applying the DGRM to the whole European market and other predictions of market growth (based on constant annual percentage growth rates) can be seen in Figures 13.1 and 13.2.

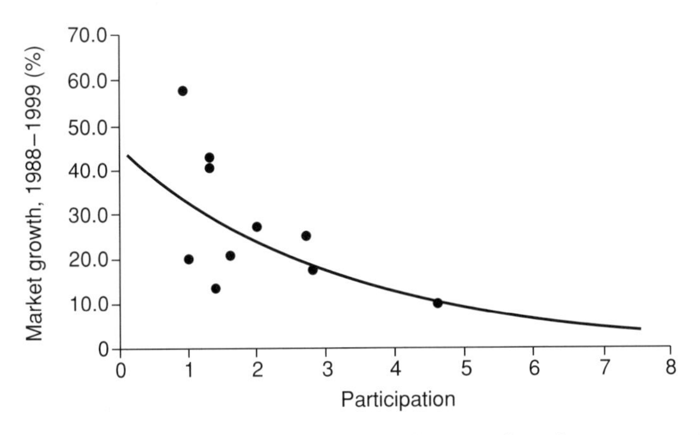

Figure 13.1 Correlation between organic retail market growth and customer participation.

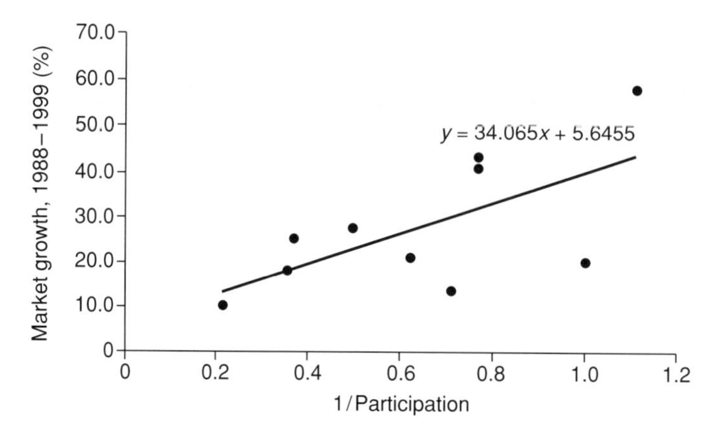

Figure 13.2 Market growth plotted against the reciprocal of participation in order to develop a linear model.

Study questions

1. Critically analyse the major factors that affect supply and demand of organic products.
2. Discuss the major characteristics of organic and conventional farming.

References

Anon. (2000) *Organic Foods in Western Europe*. Euromonitor, London.

Anon. (2001) *Next Generation Organics*. Datamonitor, London.

EC (1991) *Organic Plant Production Standards*, Regulation no. 2092/91. European Commission Publications, Brussels.

EC (1999) *Organic Livestock Production Standards*, Regulation no. 1804/99. European Commission Publications, Brussels.

Foster, C. (2000) International market growth and prospects. In *Handbook of Organic Food Processing and Production* (Wright, S. & McCrea, D.), pp. 62–77. Blackwell Science, London.

Hamm, U., Gronefeld, F. & Halpin, D. (2002) *Analysis of the European Market for Organic Food*. University of Wales School of Management and Business, Aberystwyth.

Lampkin, N. (1999) *Organic Farming*. Miller Freement, London.

Lampkin, N. & Measures, M. (2001) *2001 Organic Farm Management Handbook*. University of Wales Organic Advisory Service, Aberystwyth.

Reganold, J.P., Glover, J.D., Andrews, P.K. & Hindman, H.R. (2001) Sustainability of three apple production systems. *Nature* **410**, 926–930.

Robson, M.C., Fowler, S.M., Lampkin, N.H., Leifert, C., Leitch, M., Robinson, D., Watson, C.A. & Litterick, A.M. (2002) The agronomic and economic potential of break crops for ley/arable rotations in temperate organic agriculture. *Advances in Agronomy* **77**, 370–428.

Soil Association (1998) *The Organic Food and Farming Report 1998*. Soil Association, Bristol.

Soil Association (2000) *The Organic Food and Farming Report 2000*. Soil Association, Bristol.

Soil Association (2001) *The Organic Food and Farming Report 2001*. Soil Association, Bristol.

UKROFS (2001) *UK Organic Farm Conversion Statistics*. UK Register of Organic Food Standards, London.

Chapter 14

The US Food Supply Chain

J.R. Stock

Objectives

(1) To provide an overview of the US food supply chain.
(2) To identify the roles of various supply chain members.
(3) To identify driving forces and trends impacting the industry and individual supply chain members.

Introduction

As a staple product, consumed on a daily basis by almost 300 million persons in the United States, food is available just about everywhere. Food comes in a variety of forms such as fresh, frozen, perishable, non-perishable, processed and unprocessed, and can be purchased from a range of sources including grocery stores, vending machines, farmers' markets, military commissaries, restaurants, the Internet, mail order, and countless other outlets that sell food products in addition to their normal goods and services, e.g. cinemas, sporting events, gasoline service stations. So great is the contribution of the food industry to the US economy that in 2000 the sector contributed 12.8% of the gross domestic product (GDP) and employed more than 24 million people (Economic Research Service, 2002).

Methods of distributing food products have remained fairly traditional over the years. For the most part, food brokers, distributors and wholesalers have been key elements of the food supply chain. The channels used to distribute foods have remained static, although the uses of technology and advanced management methods have brought about significant efficiencies in the system. For the vast majority of food products (approximately 80%) truck distribution is used (McPoland, 2002). Outsourcing has become widely accepted as a means of reducing fixed assets and acquiring data and information technology. As in other industries, typical order sizes are becoming smaller, with more frequent shipments of smaller amounts gaining favour.

Overview of the US food supply chain

Historically, the food industry was highly fragmented, comprising thousands of manufacturers, wholesalers and brokers, distributing products to millions of retail outlets, food service operations, and directly to final consumers. However, since the beginning of the 20th century, an interesting phenomenon has occurred in terms of industry concentration. The US food system has developed a distribution concentration that is bimodal; i.e. larger firms have become larger and the number of smaller firms has increased (Rogers, 2001). There are many reasons for this concentration or consolidation, including the search for additional economies of scale, vertical integration in the food channel of distribution, an increasing number of contracts, partnerships and alliances between channel members, and the attempts by food manufacturers and retailers to reach an ever increasing number of diverse customer segments (Rogers, 2001; Wrigley, 2001). However, even with this trend towards concentration, for the most part the system is still made up of a very large number of small- to medium-sized companies operating locally or regionally.

As shown in Figure 14.1, the product movements within the food supply chain can take one or more different paths. In some instances, food items can go directly to the final consumer as is the case with produce being sold at farmers' markets or produce stands, or perhaps fresh fish sold at piers and the like. In other instances, food items do not go directly to final consumers, rather they go from the food source to retailers and food service organisations such as fast-food and full-service restaurants. Food may also pass through wholesalers, which is often the case with items that are sold in many grocery and convenience stores. When wholesalers sell direct to consumers, they are no longer just wholesalers, but also a type of retailer. Therefore, in Figure 14.1 there is no connecting arrow between wholesalers/ distributors and consumers.

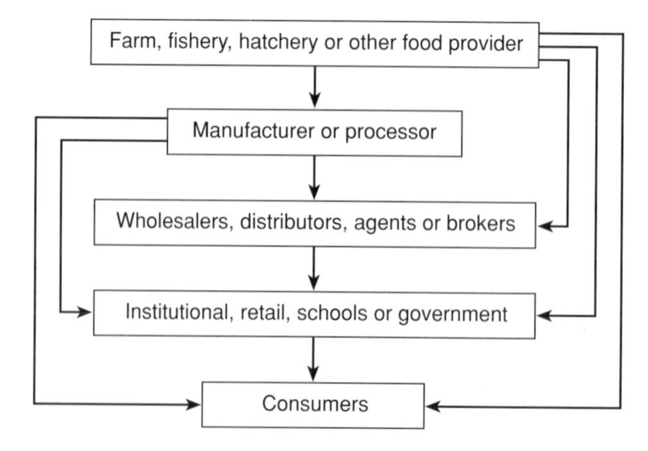

Figure 14.1 Generic overview of the US food supply chain.

For processed foods, another step is added, that of the food manufacturer. Manufacturers are typically the largest firms within food supply chains. These manufacturers are often companies very well known to consumers because of the branding of food items. Approximately two-thirds of all products sold in supermarkets or grocery stores are manufacturer brands (Marion, 1998). As is often the case, multiple channels and multiple firms are utilised, resulting in companies being associated with more than one supply chain.

Within every supply chain, and not shown in Figure 14.1, are many logistics service providers that transport, store, manage, communicate, or in other ways facilitate the movement of products and information between members of the supply chain and ultimately to final consumers. Transportation carriers, public, private and contract warehouses, logistics software vendors, returns management processors and many others, complement the physical flows that take place in the supply chain.

Finally, and also not shown in Figure 14.1, would be food importers and exporters. More than US$7 billion of food items are either imported into, or exported out of, the US each year (Economic Research Service, 2002). All members of the supply chain can, and usually do, engage in these international food distribution activities.

Roles of various members in the food supply chain

As identified in Figure 14.1, the major supply chain members will be briefly discussed, such as food providers, manufacturers/processors, wholesalers/distributors and institutions/retailers/schools/military.

Farm, fishery, hatchery and other food providers

There are thousands of farms, fisheries, hatcheries and other food providers in the USA. The largest segment of this group includes the very smallest of firms with annual sales of food products less than US$10 000 per farm (US Department of Agriculture, 1998). As mentioned earlier, there has been a concentration occurring, with many smaller farms and enterprises going out of business, merging or being acquired by larger firms. To illustrate, in 1997 there were 300 000 fewer farmers than in 1979, four firms controlled over 80% of the beef market, and even though 94% of US farms would be considered small, they received only 41% of all farm receipts (US Department of Agriculture, 1998).

Manufacturers/processors

Food manufacturers, of which there are approximately 26 000, ship products worth an estimated US$421 billion to their customers which include wholesalers, grocery stores, restaurants, convenience stores and innumerable outlets that sell or distribute food items to customers (US Census Bureau, 2001; Economic Research Service, 2002). The major components of food processing include meat products (18.5%), bakery (16.6%), fruits and vegetables (14.0%) and dairy (10.7%), with the remainder spread among sugar and confectionery, grains and oilseeds, seafood, animal foods and miscellaneous (Economic Research Service, 2002). Products from food manufacturers account for 70% of grocery store sales; ten per cent of sales would be unprocessed foods, most perishable, and the remaining 20% would be non-food products (Marion, 1998). The 100 largest firms account for three-quarters of all processed foods.

Wholesalers, distributors, agents and brokers

As reported in the 1997 Economic Census, there were 3765 wholesale grocery and related products firms. Adding the 4652 agents, brokers and commission merchants that deal in grocery products, the combined total equals more than 8000 entities (US Census Bureau, 2000). Food wholesaling is a US$589 billion business that involves merchant wholesalers, manufacturers' sales branches and offices, agents and brokers (Economic Research Service, 2002). Briefly, these forms of food wholesaling have been defined by the US Department of Agriculture's Economic Research Service as follows:

- *Merchant wholesalers*: Firms primarily engaged in purchasing groceries and related products from food processors or manufacturers and reselling them to retailers, institutions and other companies.
- *Manufacturers' sales branches and offices*: Wholesale operations maintained by grocery manufacturers or processors to market their own products.
- *Agents and brokers*: Wholesale entities that buy or sell as representatives of others for a commission and who do not usually physically handle the products.

Wholesalers may be classified as broad-line, limited-line or specialty, depending on the depth and variety of products they handle. Examples of some widely known broad-line wholesalers include firms such as Supervalue, Fleming and Nash Finch. Specialty wholesalers, handling specific product lines, account for 43% of all wholesale sales and include thousands of firms of all sizes (Economic Research Service, 2002).

Institutional, retail, school and government

It has been estimated that there were more than 158 000 grocery stores in the USA in 2001 with total sales in excess of US$517 billion (Food Marketing Institute, 2002). Grocery stores with sales in excess of US$2 million, referred to as supermarkets, total 32 265 firms with combined sales of almost US$400 billion (Food Marketing Institute, 2002). The largest 50 firms generate sales of approximately US$217 billion (US Census Bureau, 2000). About three-quarters of all grocery stores are operated by corporate chains, with most of the remaining stores part of co-operatives or voluntary groups (Marion, 1998).

Within a typical supermarket, 25 000 items are offered for sale, and inventory turns over an average of 14 times per year (all items, including non-food). Of course, inventory turnover rates vary considerably across food categories, ranging from 14 for most grocery products to more than 58 for produce. Non-food items have inventory turnover rates ranging from four to six per year (Food Marketing Institute, 2002).

Food service providers, e.g. restaurants, experienced total food and beverage spending in 1999 of almost US$900 billion (US Department of Agriculture, 2002). Of this total, approximately one-half was spent by consumers on food away-from-home. Full-service and fast-food restaurants accounted for the majority of these expenditures (MarketResearch.com Academic, 2002).

Logistics service providers

As in every sector of the economy, logistics service providers assist in the movement, storage and management of vast quantities of inventory, including transportation carriers (primarily trucks), storage facilities (public, private and contract), third parties of various kinds who administer and/or manage products or information. According to the Census Bureau, there were more than 800 000 transportation and warehousing firms in the USA with revenues of almost US$350 billion (US Census Bureau, 1997). Of this number, approximately one-half were trucking companies.

Since the 1980s, many food-related companies have participated in the growing trend to outsource logistics services to third parties. Strategic partnerships and alliances have developed which have resulted in reduced costs and improved service levels to supply chain members and their customers. Many food products now are distributed by contract carriers, many of who are strategic partners, and stored in warehouses managed by third parties.

Principal driving forces and trends affecting the food industry

There a number of driving forces behind the changes occurring in the US food industry. These forces can be grouped into four major categories: technology, regulation, environmental factors, and management orientation/philosophy.

Technology

In the technology area, radio frequency devices, bar coding, facility software management systems, such as warehouse management systems (WMS), and the Internet have been widely adopted by many firms within the food supply chain. The adoption and implementation of technology have helped shape the food supply chain as it now exists. For example, food wholesalers now do 18% of their total buying on-line, restaurants purchase almost twice as much, and food processors 14%, all via the Internet (Tompkins Associates, 2000). The integration of information systems that are electronic data interchange (EDI) or Internet based into the food supply chain is bringing about improved management of food product inventories, greater efficiency in distribution, and improved customer service levels. Much still must be done, however, before truly significant supply chain benefits are achieved (McPoland, 2002).

Adoption of technology has had a marked impact on helping to reduce one of the food industry's most difficult problems, out-of-stock products. Whether an item is unavailable from the manufacturer, wholesaler, food service provider or retailer, the result is still the same – the customer cannot make a purchase! It has been estimated that the cost of product stock-outs in the food industry is US$7–12 billion per year (McPoland, 2002).

Additionally, the improvements occurring in transportation coupled with enhancements in food preservation techniques have allowed firms to shift facility locations closer to points of food production (producer) rather than food consumption (consumer) (Rogers, 2001). Coupled with strategic partnerships and alliances between carriers, warehouses and supply chain partners, cost reductions and improved customer service levels have resulted.

Regulation

In the area of regulation, the involvement of federal agencies in food production and distribution (US Department of Agriculture, Food and Drug Administration), product liability and traceability (recalls), and security issues are important. The industry has been very concerned with health and safety issues almost from the beginning, when it was recognised that food could be contaminated by disease organisms through improper storage, handling, processing and/or preparation. In recent years, however, additional emphasis has been placed on food tampering, especially with concerns over bio-terrorism and general terrorism awareness since the 11 September 2001 attacks on the World Trade Center in New York and the Pentagon in Washington, DC. As a result, federal agencies have increased prevention, regulation and enforcement laws, policies and procedures related to product tampering and product recalls. Food safety and security have become areas of major concern by local, state and federal agencies. As an example, when DNA fragments from genetically modified maize (corn) were found in taco shells, Taco Bell, Kraft Foods and others removed the affected items from their inventories. The situation raised a number of questions about food safety and also about product traceability. Traceability in the food industry is referred to as identity preservation and is becoming a more important issue with the growth in genetically modified foods (Kuhel, 2001a).

Environmental factors

In the environmental arena are changes in the socio-economic patterns of consumers, world food needs, and the effects of pesticides and other additives on the elimination of pests and the enhanced growth of foodstuffs. Long-term changes in consumer food consumption patterns have resulted from societal developments such as the increase in the number of single parent households, multiple income families with their larger amounts of disposable personal incomes, rising importance of leisure activities coupled with a reduced emphasis on work, acceptance of Internet shopping as a viable means of purchasing food items, increasing amounts of away-from-home eating, as well as many other social and cultural changes. These changes have impacted the food industry primarily as a result of the different eating habits of consumers and the mix of food products sought by them (Economic Research Service, 2002).

World food needs are projected to grow. At the same time, at a growth rate of 1.1%, US population will double to more than 500 million during the next 60 years. Pimentel & Giampietro (1994) state that, 'If present population growth, domestic food consumption and topsoil loss trends continue, the US will most likely cease to be a food exporter by approximately 2025 because food grown in the US will be needed for domestic purposes'. While such long-range projections may be inaccurate, they do indicate that it is possible that domestic food supply chains may become less global, at least on the export side.

With increasing awareness of environmental factors, consumers have become concerned about the use of pesticides on farms and the use of genetically modified foods. With increasing governmental regulation in these and other areas impacting farming, it is likely that some elements of the food supply chain will be affected. At present, these impacts are unknown, but it would appear that they would involve the amounts of foodstuffs produced and how these products would be distributed and tracked within the supply chain.

Management orientation/philosophy

Finally, the orientations and philosophies adopted by corporations are also significant, and include such concepts as just-in-time (JIT), quick response (QR), efficient consumer response (ECR), and collaborative planning, forecasting and replenishment (CPFR). These approaches, and many others such as total quality management, process re-engineering, and continuous improvement, attempt in some way to optimise customer service and minimise logistics costs in any industry, although especially in the food supply chain.

JIT is an approach that attempts to minimise inventories while maintaining sufficient inventories of products to satisfy the needs of supply chain members and final customers. Many companies in the food industry, as well as most other business sectors, have applied JIT procedures to their operations (Stock & Lambert, 2001). QR applies JIT approaches throughout the supply chain, from food producer to consumer. 'The concept works by combining electronic data interchange (EDI) with bar-coding technology. Sales are captured immediately. This information can be passed on to the manufacturer, who can then notify its raw material suppliers and schedule production and deliveries as required to meet replenishment needs' (Stock & Lambert, 2001, p. 40).

In the retail food sector, ECR, an extension or adaptation of QR, is an important approach to managing demand at the point of sale. With changing consumer needs, firms must be more responsive in providing food products when, where and how customers want them (Hoban, 1998). The ECR strategy was developed in 1994 to improve efficiency in the following four areas (Hoffman & Mehra, 2000):

(1) Optimising store assortments and space allocations to increase category sales per square foot and inventory turnover.
(2) Streamlining the distribution of goods from the point of manufacture to the retail shelf.
(3) Reducing the cost of trade and consumer promotion.
(4) Reducing the cost of developing and introducing new products.

Combined with the use of technology, ECR can have dramatic impacts on supply chain efficiency and profitability. In the instance of a shopper buying a food product at a grocery store, as soon as the item is bar code scanned at the cash register, inventory records are updated; when the reorder point is reached, an electronic order is automatically sent to the supplier, vendor or distribution centre where the product is available. The electronic communication can take place using electronic data interchange (EDI) or via the Internet. A variation of this approach is scan-based trading (SBT) which utilises the same approach as ECR, as well as direct store delivery, except that it employs consignment delivery. Food suppliers manage their own products until they are sold and then electronic payment is made by the retailer to the supplier (Kuhel, 2001b).

CPFR, a further development of ECR, adds to ECR by stressing more collaborative relationships and information sharing (Doherty, 1998). It has been estimated that in the grocery industry supply chain alone US$150–250 billion in inventory costs could be saved with the implementation of CPFR (Doherty, 1998). 'Seamless' integration of the supply chain can result in significant cost savings. Traditionally, when information or other types of gaps occur between supply chain members, inventory is used as 'filler'. Reductions of this filler eliminate inventory carrying costs and hence give cost savings.

In addition to the above developments and trends, Marion (1998) identified several more that have impacted the food supply chain:

- *Shift in power from food manufacturers to food retailers*: The growing power of mass merchandise retailers such as WalMart, Target and Kmart, that also sell food in addition to general merchandise, as well as the growth in chain supermarkets (e.g. Krogers, Safeway, Publix) and food service providers (e.g. McDonalds, Wendy's, Taco Bell), have shifted channel power away from manufacturers. These retailers are larger in terms of assets and revenues than the manufacturers and processors that service them with products.
- *Increases in the number and scope of 'vendor participation programmes'*: Examples include co-operative advertising, in-store demonstrations, product giveaways, point-of-purchase displays, and other retailer incentives.
- *Increasing slotting fees*: In essence, these are fees charged to manufacturers to 'rent' shelf space in the retail store. With increasing numbers of new and/or expanded product lines, these fees have tended to increase, making it more

expensive for manufacturers and processors to get their products in retail grocery stores and supermarkets. Related to the increasing concentration occurring in the food industry, increases in slotting fees make it more difficult for small- and medium-sized manufacturers and processors to penetrate the retail market with their products.

In sum, there are many factors that have impacted the food industry and the food supply chain. The relationships between food product characteristics, industry drivers and trends, and customer behaviour are complex and not well understood (Hobbs & Young, 2000). However, these issues are being investigated and solutions implemented by all members of the supply chain.

Summary

While remaining somewhat traditional, the food supply chain is undergoing changes that will ultimately impact all facets of food production and distribution. The traditional enterprises involved in the food industry such as producers, manufacturers, processors, wholesalers, distributors, agents, brokers and retailers will remain, but their roles will undoubtedly change as a result of numerous trends occurring in the marketplace. The brief overview of the US food supply chain in this chapter identified the components of the supply chain and their respective activities and/or tasks. Various driving forces and trends affecting the industry have been identified and discussed relative to the production, distribution and consumption of food products. The major factors presented have included technology, regulation, environment, and management orientation/philosophy.

Study questions

1. Within the US food industry, there has been a trend towards concentration or consolidation of farms, manufacturers, retailers, etc. What are some reasons for this trend that has been occurring since the beginning of the 20th century?
2. What are the principal enterprises or types of firms that make up a typical US food supply chain? Briefly describe each of these major enterprises in terms of their position within the supply chain and their respective tasks or activities.
3. Briefly describe the various forms of food wholesaling, including merchant wholesalers, manufacturers' sales branches and offices, and agents and brokers.
4. Several driving forces or trends were identified that have impacted the US food supply chain: technology, regulation, environmental factors and management orientation/philosophy. Briefly identify and discuss the major aspects of each of these four forces or trends as they impact members of the food supply chain.

References

Doherty, K. (1998) How far to supply chain paradise? *Food Logistics* **14**, 23–32.

Economic Research Service (2002) Food market structures. *ERS/USDA Briefing Room*, United States Department of Agriculture, Washington DC (http://www.ers.usda.gov/briefing/foodmarketstructures).

Food Marketing Institute (2002) *Facts and Figures.* Washington DC (http://www.fmi.org).

Hoban, T.J. (1998) Food industry innovation: efficient consumer response. *Agribusiness* **14**, 235–245.

Hobbs, J.E. & Young, L.M. (2000) Closer vertical co-ordination in agri-food supply chains: a conceptual framework and some preliminary evidence. *Supply Chain Management: an International Journal* **5**, 131–142.

Hoffman, J.M. & Mehra, S. (2000) Efficient consumer response as a supply chain strategy for grocery businesses. *International Journal of Service Industry Management* **11**, 365–373.

Kuhel, J.S. (2001a) Going against the grain. *Supply Chain Technology News* **3**, 18–22.

Kuhel, J.S. (2001b) Scan-based trading. *Supply Chain Technology News* **3**, 37–39.

McPoland, D. (2002) Leading trends that will impact the food supply chain over the next five years. Logistics and supply chain site (http://www.logistics.about.com).

Marion, B.W. (1998) Changing power relationships in US food industry: brokerage arrangements for private label products. *Agribusiness* **14**, 85–93.

MarketResearch.com Academic (2002) *Foodservice Industry* (http://www.academic.marketresearch.com).

Pimentel, D. & Giampietro, M. (1994) *Food, Land, Population and the US economy.* Carrying Capacity Network, Washington DC (http://www.dieoff.com).

Rogers, R.T. (2001) Structural change in US food manufacturing, 1958–1997. *Agribusiness* **17**, 3–32.

Stock, J.R. & Lambert, D.M. (2001) *Strategic Logistics Management*, 4th edn. Irwin/McGraw-Hill, Boston.

Tompkins Associates (2000) Food industry e-commerce takes off. *Competitive Edge* **2**, 9.

US Census Bureau (1997) *Non-employer Statistics: Transportation and Warehousing*. Economic Census Report, Washington DC (http://www.census.gov/epcd/nonemployer/1997).

US Census Bureau (2000) *Concentration Ratios in Retail Trade 1997.* US Government Printing Office, Washington DC.

US Census Bureau (2001) *Concentration Ratios in Manufacturing 1997.* US Government Printing Office, Washington DC.

US Department of Agriculture (1998) *A Time to Act.* US Government Printing Office, Washington DC (http://www.usda.gov).

Wrigley, N. (2001) The consolidation wave in US food retailing: a European perspective. *Agribusiness* **17**, 489–513.

Chapter 15

The Future of Food Supply Chain Management

C. Bourlakis & M. Bourlakis

Objectives

(1) To discuss a range of issues likely to affect the evolution of future supply chains.
(2) To make researchers, managers, students and policymakers alert to those factors that have the potential to assert an influential impact upon the strategic development of the food supply chain in the near future.

Introduction

The field of food supply chain management has undergone tremendous changes over the past 35 years. Food supply chain management that was once considered the last frontier of cost reduction in the 20th century has now become the major strategic issue for firms in the new millennium. One of the main goals of the food firm is to establish a combination of purchasing, transportation, physical distribution and logistics that will put the company in a position to achieve economies of scale.

This chapter aims to analyse a wide spectrum of issues that will influence the evolution of strategic development of future food supply chains. We acknowledge the fact that it is almost impossible to incorporate and to envisage the full range of issues related to such an evolving field; however, we anticipate that our analysis will shed some light on future trends and developments taking place in the food supply chain management field.

Evolution of the food sector in the 20th century

According to Ramsey (2000) there were four distinct phases in the evolution of the food sector in the 20th century:

(1) The era of early competition (1900–1920).
(2) National consolidation (1920–1945).

(3) Internationalisation (1945–1980).
(4) Globalisation (1980–2000).

During the era of early competition that took place between 1900 and 1920, the supply chain was in the hands of regional wholesalers, whereas during the national consolidation period from 1920 to 1945, national manufacturers with the support of the wholesalers and retailers ran the supply chain. The third period of the 20th century from 1945 to 1980 saw the expansion of many leading food companies from their home market to the international competitive arena, during which the supply chain was dominated by food manufacturers who now faced some retail 'countervailing' challenge as a result of increased concentration in the retail sector. Finally, during 1980–2000, the so called globalisation phase, supply chain control shifted from manufacturers to retailers in Europe, while retailers were challenging for dominance over manufacturers in North America (Ramsey, 2000). We are currently witnessing an increasing rate of globalisation, with European supermarket retailers and US food manufacturers taking the lead (Ramsey, 2000); following suit are the third party logistics firms which are gradually engaging in the globalisation process (Stone, 2001).

Increase of logistics externalisation

Third party logistics, contract logistics, externalisation and outsourcing generally mean logistics services provided by a single vendor on a contractual basis. Logistics externalisation can contribute in raising a firm's competitive position by enhancing customer service, facilitating the opening of new markets and, most importantly, providing the company that chooses such an option with dedicated resources. Fernie (1989) provided a list of the beneficial reasons for the use of either third party or own-account distributors. These reasons were classified into three categories for contract distribution – strategic, financial and operational – whereas the own-account distribution incorporates the aspects of cost, control and economies of scale, as listed in Box 15.1. Overall, the decision that matters is whether to externalise (contract out) or to internalise (own-account) the distribution functions. Fernie (1990) argued that the element of cost-effectiveness is the most important one when taking the relevant decision, although whatever option is taken, some management expertise will still have to be retained to monitor the performance of the contractor.

At a European level, the deregulation of road haulage that has been occurring over the past 30 years in most European countries has made it much easier and more attractive for firms to contract out the transport function. However, McKinnon (1998) argues that the recent growth in the externalisation of that function is not due exclusively to the deregulation of road haulage, but to other factors as well. For example, in the UK, this was attributed to a general change in managerial attitudes to contracting out that took place between the 1980s and the 1990s. Overall, this trend has been developed to a high degree within the western European environment,

Box 15.1 The benefits of contract vs. own-account distribution (Fernie, 1989).

Contract distribution	Own-account distribution
Strategic reasons	*Cost*
Flexibility	Cost plus argument
Spread risks	Monitoring costs
Financial reasons	*Control*
Off balance sheet financing	Total responsibility in the supply chain
Opportunity cost of capital investment	Better customer service
Better planned budgets	Loyalty to one, not several companies
Operational reasons	Security for new product development
Accommodate seasonal peaks	Economies of scale
Reduce backdoor congestion at warehouse/store	'In-house' technological innovation
Provision of 'specialist services'	
Improve service levels	
Management expertise	
Minimise industrial relations problems	

but for the rest of Europe many differences across countries and regions still exist. For example, in southern Europe the level of third party penetration is far smaller than in western Europe (Bence, 1995).

Hence, there is great potential for logistics externalisation, given that in 1996 just under one-quarter of total logistics expenditure in the European Union was outsourced, with a considerable amount of variation between various EU members. For example, UK firms externalised around 34% of their logistics spending, whereas in Greece only 11% was contracted out (Marketline International, 1997). Most surveys on the issue forecast an increase in logistics and supply chain externalisation at the European level (McKinnon, 1999).

A survey of 300 food and non-food European firms (manufacturers, retailers, etc.) identified the factors that affect a firm's choice for selecting distribution service providers (P-E Consulting, 1996). These were, in order of importance: ability to provide national transport services, EDI communication, ability to provide international transport services, ability to provide services on a pan-European basis, and the provision of warehousing and other dedicated services. According to that survey, although these third party firms may offer a satisfactory service at a national level, their services were considered to be less satisfactory on a pan-European level, connoting the future growth potential for logistics externalisation in that region (P-E Consulting, 1996).

Formation of fourth party networks

The increased pressure for higher cost efficiency affects all members of the food chain; for example in the USA, fruit and vegetable chain retailers demand greater efficiency in distribution from their suppliers (Epperson & Estes, 1999), and this is also the case in other countries. As a consequence, the food chain has moved from a fragmented and decentralised structure to a more vertically integrated structure that includes joint partnerships, strategic alliances, and in general there are now more vertical co-ordination agreements among food chain members (Epperson & Estes, 1999). Thus new value-added logistics services (e.g. warehouse and fleet management, information technology and Internet software development, route and product planning) have been demanded by retailers (McKinnon, 1999). Traditional third party logistics firms could not fully match the demands of the retailers, because they did not possess the appropriate logistics and information technology expertise. In many cases, UK food multiple retailers, such as Tesco and Sainsbury, opted to co-operate with logistics information technology suppliers such as the consulting firm Accenture. A company like this can assist the development of logistical services with the close collaboration of both retailers and their traditional logistics partners, the so-called third party logistics companies. The new range of services (value-added services) adds additional operational complexity to the food retail chain, because increasingly specialised logistics information technology firms can be involved.

In the authors' opinion, a fourth party logistics network (4PL™ by Accenture) needs to be created to take control of all aspects of the retailer's logistics operations, but in close collaboration with the traditional third party logistics firms and the logistics information technology suppliers. The 4PL firm in command of the network would act as a channel integrator and an interface between the retailer, its traditional third party logistics firms and logistics information technology suppliers. It is expected to be an independent management company whose role will be to provide supply chain services to a primary client and, at a later stage, to provide services to other clients in related and unrelated industries (Gattorna, 1998). The 4PL network benefits largely from the recent information technology advances, most notably the Internet, that support the effective synchronisation and cost-efficient co-ordination among the chain members, and has the potential to become the most efficient organisational mode in that type of operation. The 4PL network co-ordinator will also be in a position to reduce the transaction costs incurred as a result of new services required by the retailer, and the numerous inter-organisational links created by the retailer and its various logistics services providers.

Agility in the food supply chain

In today's and tomorrow's volatile markets, more demand- and customer-driven chains are becoming essential for company survival as product life cycles shorten due to increased competition. The food supply chain of the future should be agile enough to respond to consumer demand. We must distinguish and not confuse the difference between an agile and a lean supply chain (Christopher, 2000). Leanness in production entails an element of technical efficiency, where maximum production is achieved with the least use of inputs, and with the term being associated with lean manufacturing where the product is delivered in a just-in-time fashion. Leanness in production functions best in low variety and high volume production environments, such as the cement industry. Agility in production seems to be a very important factor in the food supply chain because market sensitivity is absolutely essential in the food industry. Agility is required in business environments where demand is more volatile, less predictable, and the need for product variety is high (Christopher, 2000). Agility in production is basically the ability to be market sensitive by understanding and responding to customers' needs. Agility in supply chain management entails a number of characteristics that embrace logistic processes, the firm's information system infrastructure, the type of relationship that the company has with its suppliers, and other elements referring to its overall organisational structure. However, the underlying element of agility is capability of flexible production.

Sustainable integration and co-ordination in the food supply chain

The trade-off of one cost against another has been the main task of product flow management since the early 1960s, where the optimisation of conflicting product flow activities contributed significantly to reduction in costs. Cost conflicts could arise in areas such as production, purchasing, transportation and warehousing. This type of intrafunctional supply chain co-ordination comes under the auspices of product flow management and is regarded as an efficient route to cost control. However, the management of logistical activities in the supply chain involves other functions within the company, such as production, marketing and finance. By changing the level of the various logistical activities, such as transportation and warehousing, it is very likely that the firm will affect the targets set by other functions, such as marketing and finance. In such a case it is essential for the firm to achieve interfunctional co-ordination by balancing the effect of the change with those of other functional areas.

Ballou *et al.* (2000) argue that supply chain management has now moved from an intrafunctional vision of the channel towards an interfunctional and even an interorganisational one. Following Ballou *et al.* (2000), cost service improvement may be found in co-ordinating product flows among a number of firms, as the current task of management is to seek lower costs beyond the boundaries of their own enterprises, as for example with the use of a third party logistics company. The latter requires the co-ordination of all channel members involved; a prerequisite that the coalition will remain solid is that no member of the inter-organisational co-ordination benefits at the expense of others. Ballou *et al.* (2000) point out that there appear to be three elements that are essential to the inter-organisational management for the supply chain coalition to remain stable:

(1) A new inter-organisational type of metrics is required to define and measure costs among the various channel members.
(2) An information sharing mechanism must be established to convey information about the benefits of co-operation among the channel members.
(3) The allocation of benefits must be distributed fairly to all channel members, i.e. a proper and just allocation method for redistributing the benefits of co-operation must be established.

Green and reverse supply chains

In future, companies will be invited to reduce still further their adverse environmental impact, so the 'greening' of the supply chain and the recycling of products will increasingly come into societal and managerial focus. Reverse logistics is a concept that is applied in processes associated with recycling, reducing and reusing the materials used in the production process. The process can start as early as the product development phase where companies consider which input materials to use in their product to minimise input costs and also the costs of separating and recycling them at a later stage after consumption of the product and their return. Although for some manufacturing products and services, such as printers, cars, chemicals and packaging, recycling systems are already in place, for other products the recycling process has not yet been systematised, but the relevant process is just starting. It appears that reverse logistics will be an important item on the cost minimisation agenda in the future management of the food supply chain; it may also be the case that reverse logistics, if used effectively, can become a source of competitive advantage. Van Hoek (1999) put forward a number of propositions for turning the current state of reversed logistics into the green supply chains of the future, via an active greening involvement of all upstream, midstream and downstream players throughout the supply chain.

Another environmental concern is the creation of 'food miles'. Food miles represent the distance food travels from where it is grown or raised to where it is ultimately purchased by the consumer or end user (Pirog *et al.*, 2001). If a consumer is purchasing foreign produce that can be simultaneously produced at domestic level, the former represents a 'high food miles product' compared to the latter. Moreover, we need to consider the environmental and resource implications of food miles, such as pollution and congestion, together with animal welfare issues for livestock produce. For example in the UK red meat chain, sheep from northern regions are typically sent to abattoirs in the south for slaughter, only for the carcasses to be returned to regional distribution centres and stores in the north (Bourlakis & Allinson, 2002). Unavoidably, such animal transportation creates extra food miles and is arguably unsustainable in terms of cost efficiency and animal welfare. There is a need for shorter food supply chains, which can be achieved if retailers increase their purchasing of local and regional food produce.

Collaboration and partnerships between members of the food chain

The world food industry is a highly segmented market in which the changing nature of consumer preferences and the need to respond to a specific country's consumer tastes mean that food manufacturers and retailers need to be able to respond rapidly to developments. Although mergers and acquisitions have already played a significant part in reshaping the world food industry, it is expected that horizontal strategic alliances will continue to provide a more flexible way of increasing market share throughout the food chain, especially for food manufacturers. For example, two manufacturers of food products can combine their fleet of trucks to deliver products to each other, and by so doing save substantial sums which would otherwise have been spent on product distribution. Alliances between two food manufacturing firms may also be able to exploit economies of scope in production, as when a food company undertakes to produce in one of its plants some products and brands of another food manufacturing company belonging to a horizontal alliance, and then delivering these to the nearest consumer market. Horizontal alliances also occur among food retailers in the form of joint retail bulk buying with the aim of increasing their bargaining strength when dealing with powerful food manufacturing firms.

Fearne (1994) proposed that in the European food industry vertical co-ordination in supply chain management in the form of alliances and partnerships would increase substantially for the following reasons:

(1) The incentive provided to all parties concerned to reduce costs in the supply chain and to increase their competitiveness.
(2) The substantial rise in the growth of 'own label' retail food products, alongside the decrease in the life-cycles for food products as consumer demands change rapidly.
(3) The increased demand for products that have a shorter shelf life, such as fresh and chilled food products.
(4) The growing concern and the increased awareness of consumers over how, where and by whom food products are produced.
(5) Food safety legislation.

A recent study by Fearne & Hughes (2000) has looked into the transformations that have taken place over the past decade in the UK fresh produce industry and identified four key factors that have played a significant part in the transformation of the UK towards a culture of value creation and innovation:

(1) The strategies followed by supermarkets operating in the UK.
(2) Supply chain integrity and food safety legislation.
(3) Innovation.
(4) Rationalisation of the supply base.

The major four multiples in the UK (Tesco, Sainsbury, Asda and Safeway) now focus their strategies on product differentiation and at the heart of the growth strategies there should be the development and promotion of own label (including meat and fresh produce) products, rather than price competitiveness and the large product range approaches seen in the past.

The increased presence of own label products in supermarkets, together with the 1990 Food Safety Act requiring retailers to ensure that the food they purchase from upstream producers is safe, led to multiples increasing their involvement in upstream supply chain management (Fearne & Hughes, 2000). However, the attempts by multiples in the UK to ensure greater supply integrity and quality assurance of fresh produce had to be matched by greater control of the supply chain; this lead to endeavours to rationalise the supply base and minimise costs by dealing with fewer food suppliers that were more efficient and ready to respond to innovations. Fearne & Hughes (2000) concluded that greater shifts towards collaboration and supply chain partnerships, where greater emphasis will be placed upon return on investment and value added to all participants, will ensue. A study by White (2000) on buyer–supplier relationships for the UK fresh fruit and vegetable industry confirmed that trading between multiple retailers and their fruit and vegetable suppliers ceased to be purely transactional about ten years ago, and suggested that towards the end of the 1990s the relationship between the two parties became more co-operative with the multiples less exploitative of their suppliers.

The removal of transport and trade barriers between EU countries, the opening of markets in eastern Europe and the emergence of pan-European logistics service providers will facilitate further the growth of strategic supply chain partnerships within Europe.

Role of the internet in the food supply chain

The rapid development of information technology and communication systems will enhance the globalisation of the supply chain, and it is expected that logistics practices will be dominated by e-commerce, i.e. internet-supported activity coupled with strategic partnerships and globalisation, in the supply chain (Skjoett-Larsen, 2000). In addition, the internet provides a solid platform for conducting business in a unique way, where all supply chain members interact with each other (Stough, 2001). In this context, the internet can support a customer-focused and customer-responsive organisation; more specifically, the internet offers firms a number of strategic, selling and cost-cutting opportunities in the food supply chain such as the following (see Lancioni *et al.*, 2000):

(1) On-line vendor catalogues, from which customers can select and order food items from retailers.
(2) The efficient processing of customer complaints and queries, and the timely receipt of orders from customers in other countries.
(3) Tracking shipments transported via rail, air or truck, and the scheduling of pick-ups and deliveries.
(4) The provision of 24 hours a day/7 days a week customer service, and communication with suppliers on a worldwide basis.

It can therefore be reasonably suggested that use of the internet will be far more extensive in future food supply chains, provided its use gives firms a competitive advantage.

Conclusions

The key issues that could engineer changes in the food supply chain have been discussed. It is suggested, regarding the externalisation of supply chain operations, that specialist logistics firms will have a more central role to play in future, especially when they are able to offer a full range of services. Greater use of the internet is another expected trend. Environmentally conscious consumers will also exert a large influence on future food supply chains. It is also anticipated that food channel members (manufacturers, retailers and logistics firms) will form networks, and the potential emergence of a fourth party logistics network commander has been discussed. Finally, it is forseen that the factors described above will increase competition at retail and manufacturing levels to the benefit of European consumers' welfare, in terms of both lower product prices and safer and higher quality products.

Study questions

1. Analyse some of the key issues that may have an influential impact on the evolution and strategic development of food supply chains in the future.
2. With reference to logistics externalisation, discuss and classify the benefits of contract versus own account distribution and analyse the future potential of that process at the European level.

References

Ballou, R.H., Gilbert, S.M. & Mukherjee, A. (2000) New managerial challenges from supply chain opportunities. *Industrial Marketing Management* **29**, 7–18.

Bence, V. (1995) The changing marketplace for distribution: an operator's perspective. *European Management Journal* **13**(2), 218–229.

Bourlakis, M. & Allinson, J. (2002) *The Aftermath of the Foot and Mouth Crisis in Agricultural Logistics: The Case of the UK Fat Lamb Chain*. Research Study sponsored by the Institute of Logistics and Transport, Institute of Logistics and Transport Research Series 2001/2, The Centre of Rural Economy, School of Agriculture, Food and Rural Development, University of Newcastle upon Tyne, UK.

Christopher, M. (2000) The agile supply chain – competing in volatile markets. *Industrial Marketing Management* **29**, 37–44.

Epperson, J.E. & Estes, E.A. (1999) Fruit and vegetable supply-chain management, innovations, and competitiveness: co-operative regional research project S-222, *Journal of Food Distribution Research* **30**(3), 38–43.

Fearne, A. (1994) Strategic alliances in the European food industry. *European Business Review* **94**(4), 30–36.

Fearne, A. & Hughes, D. (2000) Success factors in the fresh produce supply chain – insights from the UK. *British Food Journal* **102**(10), 760–772.

Fernie, J. (1989) Contract distribution in multiple retailing. *International Journal of Physical Distribution and Materials Management* **19**(7), 1–35.

Fernie, J. (1990) Third party or own account – trends in retail distribution. In *Retail Distribution Management* (Fernie, J., ed.), pp. 91–106. Kogan Page, London.

Gattorna, J. (1998) Fourth party logistics: en route to breakthrough performance in the supply chain. In *Strategic Supply Chain Alignment* (Gattorna, J., ed.), pp. 425–445. Gower, Aldershot.

Lancioni, R.A., Smith, M.F. & Oliver, T.A. (2000) The role of the internet in supply chain management, *Industrial Marketing Management* **29**, 45–56.

Marketline International (1997) *EU Logistics*. Marketline International, London.

McKinnon, A. (1998) The abolition of quantitative controls on road freight transport. *Transport Logistics*, **1**(3), 211–224.

McKinnon, A.C. (1999) The outsourcing of logistical activities. In *Global Logistics and Distribution Planning* (Walters, D., ed.), pp. 214–239. Kogan Page, London.

P-E Consulting (1996) *Logistics in Europe: The Vision and the Reality*. P-E Consulting, Surrey.

Pirog, R., Van Pelt, I., Enshayan, K. & Cook, E. (2001) *Food, Fuel and Freeways: An Iowa Perspective on How Far Food Travels, Fuel Usage and Greenhouse Gas Emissions*. Leopold Center for Sustainable Agriculture, Iowa State University, IA.

Ramsey, B. (ed.) (2000) *The Global Food Industry: Strategic Directions*, Financial Times Retail and Consumer Publishing, London.

Skjoett-Larsen, T. (2000) European logistics beyond 2000. *International Journal of Physical Distribution and Logistics Management* **30**(5), 337–387.

Stone, M.A. (2001) European expansion of UK third party logistics service providers. *International Journal of Logistics: Research and Applications* **4**(1), 97–115.

Stough, R.R. (2001) New technologies in logistics management. In: *Handbook of Logistics and Supply Chain Management* (Brewer, A.M., Button, K.J. & Hensher, D.A., eds), pp. 513–520. Pergamon-Elsevier, Oxford.

Van Hoek, R.I. (1999) From reversed logistics to green supply chains. *Supply Chain Management: an International Journal* **4**(3), 129–134.

White, H.M.F. (2000) Buyer–supplier relationships in the UK fresh produce industry. *British Food Journal* **102**(1), 6–17.

Index

Page numbers in *italic* refer to figures, tables and boxes. Abbreviations used in the index are: BSE = bovine spongiform encepalopathy; CAP = common agricultural policy; CJD = Creutzfeldt-Jakob disease; CPFR = collaborative planning, forecasting and replenishment; ECR = efficient consumer response; HACCP = Hazard Analysis Critical Control Point; IT = information technology; TCSCs = temperature controlled supply chains.